ISO 9000

Manufacturing, Software, and Service

Charles A. Schuler
Professor, California University of Pennsylvania
Department of Industry and Technology

Jesse Dunlap
President, Jesse Dunlap Associates

Katharine L. Schuler
Technical Writer, ABB Automated Distribution Division

Delmar Publishers

1945 - 1995
50 years

I⬥P™ An International Thomson Publishing Company

Albany • Bonn • Boston • Cincinnati • Detroit • London • Madrid •
Melbourne • Mexico City • New York • Pacific Grove • Paris •
San Francisco • Singapore • Tokyo • Toronto • Washington

NOTICE TO THE READER

Cover photo courtesy of NASA
Cover Design: Steele Graphics

Delmar Staff
Publisher: Robert Lynch
Administrative Editor: John Anderson
Senior Project Editor: Christopher Chien
Production Manager: Larry Main
Art and Design Coordinator: Nicole Reamer
Editorial Assistant: John Fisher

COPYRIGHT © 1996
By Delmar Publishers
a division of International Thomson Publishing Inc.

The ITP logo is a trademark under license.

Printed in the United States of America

For more information, contact:

Delmar Publishers
3 Columbia Circle, Box 15015
Albany, New York 12212-5015

International Thomson Publishing
Europe
Berkshire House 168-173
High Holborn
London, WC1V 7AA
England

Thomas Nelson Australia
102 Dodds Street
South Melbourne, 3205
Victoria, Australia

Nelson Canada
1120 Birchmont Road
Scarborough, Ontario
Canada, M1K 5G4

International Thomson Editores
Campos Eliseos 385, Piso 7
Col Polanco
11560 Mexico D F Mexico

International Thomson Publishing GmbH
Konigswinterer Strasse 418
53227 Bonn
Germany

International Thomson Publishing Asia
221 Henderson Road
#05-10 Henderson Building
Singapore 0315

International Thomson Publishing—Japan
Hirakawacho Kyowa Building, 3F
2-2-1 Hirakawacho
Chiyoda-ku, Tokyo 102
Japan

2 3 4 5 6 7 8 9 10 XXX 02 01 00 99 98 97 96

Library of Congress Cataloging-in-Publication Data

Schuler, Charles A.
  ISO 9000 : manufacturing, software, and service / Charles Schuler,
Jesse Dunlap, Katie Schuler.
    p. cm.
  Includes index.
  ISBN 0-8278-7124-1
  1. ISO 9000 Series Standards. I. Dunlap, Jesse. II. Title.
TS156.6.S38 1996
658.6'62—dc20   95-22157
                                              CIP

Table
of
Contents

Preface   *xiii*

1   Introduction and History of Standards   1

Introduction   1

History   4

The ISO Standard   6

Chapter Review Questions   9

2   Benefits of ISO 9000 Registration   11

Introduction   11

Benefits to the Customer   11

Benefits to the Subcontractor   15

Benefits to the Supplier   18

Chapter Review Questions   20

3   Organizing for ISO Implementation   23

Introduction   23

Choosing the Coordinator   25

Choosing the Team   27

Financing the Effort   30

Changing Priorities   30

Chapter Review Questions   31

4   Consultants and Registrars   33

Introduction   33

Consultants   33

Registrars   37

Chapter Review Questions   42

5   *The ISO 9001 Elements*   45

Introduction   45

The Role of Leadership   45

    4.1   Management Responsibility   45

        Quality Policy   46

        Organization   47

        Resources   48

        Management Representative   48

        Management Review   49

    4.2   Quality System   49

    4.18   Training   51

Control of External Interfaces   55

    4.3   Contract Review   55

        Cross Functional Contributions   55

        Modifications to Contracts   56

    4.6   Purchasing   56

    4.7   Control of Customer Supplied Product   58

    4.19   Servicing   58

Design Control   60

    Design Philosophy   61

Document and Data Control   66

    4.5   Document Control   66

        Includes Software   67

        The Requirements   68

        Keys to Implementation   68

        Procedure for Control   69

        Method of Control   69

        Security   70

        No "Post-Its" Allowed   70

        Changes to Controlled Documents   70

        A Disciplined Approach   71

4.16 Quality Records   71

Operations Control   72

4.8   Product Identification and Traceability   72

Traceability   73

4.9   Process Control   73

Manufacturing Hub   74

Typical Audit   74

Production Equipment Too   76

Records   76

Special Processes   76

4.10 Inspection and Testing   77

Receiving Inspection   77

Inspection is Necessary   78

In-Process Inspection   78

Final Inspection   79

Inspection and Test Results   79

4.11 Inspection, Measuring and Test Equipment   79

Comprehensive List   79

Traceable Standards   80

Identification   80

Measurement Capability   81

Recall System   81

Secure Environment   82

Test Result Validity   82

Software Too   82

4.12 Inspection and Test Status   82

4.13 Control of Nonconforming Product   83

Identification   84

Documentation   84

Evaluation   85

Segregation   85

Disposition   85

Notification   86

4.15 Handling, Storage, Packaging, Preservation, and
        Delivery   86

            Final Product   87

            Expiration Dates   87

            Storage and Shipping   87

            Packaging   87

    Continuous Improvement   88

        4.14 Corrective and Preventive Action   88

            Customer Complaints   88

            Processing Problems   89

            Supplier Problems   89

            Systemic Approach   90

            Follow-up System   90

        4.17 Internal Quality Audits   91

            Formal System   91

            Training   92

            Independence   92

            Costs   92

            Records   93

            Audit Records   93

        4.20 Statistical Techniques   93

            Seven Basic SPC Tools   93

            Use It Correctly   94

            The Proper Way   94

            Recalculate Control Limits   96

    Chapter Review Questions   96

6   Relationship of ISO 9000 to TQM, MRP, and JIT   101

    Introduction   101

        ISO 9000 plus TQM equals world class quality   101

    TQM   101

        The Evolution of TQM   102

        Culture Changes   103

No Conflict   103

Relationship of ISO to TQM   104

ISO 9000 — A Full Service Business System   106

Complementary Systems   106

Enhancement of SPC   107

Document Control Example   108

MRP   109

Relationship of MRP and ISO   111

JIT   112

Push Systems and Pull Systems   113

Fat Hides Problems   115

Forming Partnerships with Suppliers   117

ISO and JIT   118

Chapter Review Questions   118

7 **_ISO 9000-3: Quality in Software_   _121_**

Introduction   121

Emerging Standards   121

The TickIt Program   123

SQSR   124

Quality System Framework   124

4.1 Management Responsibility   124

Purchaser's Management Responsibility   125

Joint Reviews   125

4.2 Quality system   126

Quality System Documentation   126

Quality Plan   127

Internal Quality System Audits   128

Corrective Action   128

Quality System Life Cycle Activities   128

5.2 Contract Review   129

Contract Items on Quality   130

5.3 Purchaser's Requirements Specification   130

       Mutual Cooperation   133

5.4 Development Planning   133

       Development Plan   133

       Progress Control   136

       Input to Development Phases   136

       Output From Development Phases   136

       Verification of Each Phase   137

5.5 Quality Planning   137

       Quality Plan Content   137

5.6 Design and Implementation   138

       Design   138

       Implementation   139

       Reviews   140

5.7 Testing and Validation   142

       Test Planning   142

       Testing   143

       Validation   143

       Field Testing   143

5.8 Acceptance   144

       Acceptance Test Planning   144

5.9 Replication, Delivery, and Installation   144

       Replication   145

       Delivery   145

       Installation   145

5.10 Maintenance   146

       Maintenance Plan   146

       Release Procedures   147

Quality System Supporting Activities   148

6.1 Configuration Management   148

       Configuration Management Plan   149

       Configuration Management Activities   149

6.2 Document Control   150

   Types of Documents   150

   Document Approval and Use   151

   Document Changes   151

6.3 Quality Records   151

6.4 Measurement   151

   Product Measurement   151

   Process Measurement   152

6.5 Rules, Practices, and Conventions   153

6.6 Tools and Techniques   153

6.7 Purchasing   154

   Assessment of Subcontractors   154

   Validation of Purchased Product   154

   Included Software Product   155

6.9 Training   155

Chapter Review Questions   156

8 ISO 9004-2: Quality in Service   159

Introduction   159

Service Quality Design   160

   Market Research   161

   Service Brief   161

   Design Changes   164

   Service Changes   164

Management Functions   165

   Management Review   169

Quality Assessment   169

   Customer Input   172

Documentation and Training   172

   Training   174

Chapter Review Questions   175

9 Implementation and Audit Preparation   177

Introduction   177

Tools   177
>   Baseline Audit   178
>   Project Plan   178
>   Corrective Action   179
>   Monitoring Progress   179
>   ISO Coordinators   180
>   Discipline   181
>   Cleanup   181
>   Registration Audit   181
>   Audit Management   182
>   Chapter Review Questions   184

10   Registration Maintenance and the Future   185

    Introduction   185
    Surveillance Audits   185
>   Weak Areas   185
    Internal Audits   186
>   Corrective Actions   186
>   Management Reviews   187
>   Other Areas   187
    Maintenance Audits   188
    The Future of Standards   188
    Chapter Review Questions   189

Appendixes   191

    A   Bibliography   191
    B   Quality Vocabulary   193
    C   Common Acronyms   201
    D   Forms and Checklists   207
>   Approved Commodity and Supplier List   208
>   Approved Supplier List   209
>   Authorization to Release Material for Urgent Production   210
>   Calibration Master Inventory   211
>   Calibration Service Record   212

Contract Quality Requirements Analysis   213

Controlled Document Inventory   214

Corrective Action Request Form   215

Design Protocol Checklist   216

Design Quality Review   217

Document Change Alert   218

Internal Audit Report   219

Internal Audit Schedule   220

ISO 9001 Review Checklist   221

Notification of Noncompliance   222

Notification of Software Contract Review   223

Software Contract Review Report   224

Software Inspection and Formal Review Error List   225

Software Inspection and Formal Review Summary   226

Software Maintenance Report   227

Subcontractor Quality Assessment   228

Supplier Ratings   229

Temporary Change Form   230

Test Data Record   231

Training Matrix   232

Training Request Form   233

Vendor Corrective Action Request   234

E   Stamps and Tags   235

Quality Stamp System   236

Source Inspection   237

Test Acceptance   237

Scrap   237

Limited Shelf Life   238

Static Sensitive   238

Temperature Sensitive   238

Calibration Due   239

Not Calibrated   239

Out of Service   239

No Calibration Required   239

Lot Traveler   240

Index *241*

Preface

This book is for everyone involved with or interested in the quality of products and services. It will serve as a textbook for college students in disciplines such as business, engineering, manufacturing, computer science, and technology. It will serve as a resource for workers and managers in companies interested in quality management and the international standards for quality.

The world-wide quality movement is in full swing. Competition and customers both demand that companies provide high-quality products and services and that they do it consistently. There have been numerous quality movements and numerous quality standards. ISO 9000 represents the first significant international standard for quality.

This book is practical and has been arranged both for readability and for referencing. Chapters 5 and 7 have italic headings with ISO element numbers. These facilitate using this book as a reference. The elements are not necessarily ordered numerically. They are grouped logically to facilitate readability. The chapters end with questions that can serve as assignments, as a self-check, or to stimulate discussions and activities when companies are engaged in the organization or implementation phases of their quality systems.

Delmar Publishers' Online Services
To access Delmar on the World Wide Web, point your browser to:
http://www.delmar.com/delmar.html
To access through Gopher: **gopher://gopher.delmar.com**
(Delmar Onlineis part of "thomson.com", an Internet site with
information on more than 30 publishers of the
International Thomson Publishing organization.)
For more information on our products and services:
email: **info @ delmar.com**
or call **800-347-7707**

1

INTRODUCTION AND HISTORY OF STANDARDS

INTRODUCTION

After World War II, the United States of America emerged as the preeminent producer of goods. The U.S. produced a large share of goods to meet the world's needs in areas such as steel, automobiles, electronics, textiles, optics, appliances, machinery, tools, and aircraft. This significant industrial output kept imports at a reasonable level, produced exports, and thus provided a healthy balance of trade with other nations. Now, however, that promise of prosperity, which is linked with strong industrial output and a reasonable balance of trade, is in jeopardy for many U.S. citizens. In some years, trade deficits have exceeded 150 billion dollars.

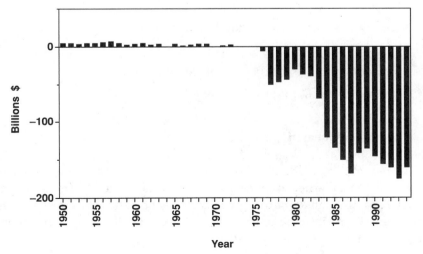

Figure 1.1 U.S. balance of trade

How has the economic outlook shifted so drastically since the 1950s and 1960s? There are many theories and, of course, the usual controversies. The viewpoint taken here is that the postwar era saw a strong U.S. economy with heavy consumer demands for goods of all types. U.S. companies, in an attempt to meet that demand, concentrated mainly on increasing production and profits. This worked, but only for briefly. In the meantime, wartorn countries such as Japan and Germany were rebuilding and taking advantage of the quality concepts that were beginning to emerge. They also incorporated newer technology and eventually, their output increased. At the same time, these countries began to produce better quality goods at lower costs than those of their American competitors.

American consumers are no different than other world consumers in the sense that two factors strongly influence most buying decisions. Those factors are *value* and *quality*. Asking a family to purchase something that costs more, does not look as good, does not work as well, and that is not as reliable—all in the name of saving some job for some unknown worker in some unknown part of the country—is simply asking too much.

The economic ills of the United States are blamed on many sources:

- unions
- management
- low U.S. import tariffs
- high foreign import tariffs
- wages
- taxes
- lack of a government-industry policy and cooperation
- cost of financing
- environmental and other government regulations
- health care costs and other benefits
- litigation
- an undereducated workforce

Although these concerns are real in the sense that they limit U.S. competition, most other nations have the same or a similar set of constraints. Also, the danger of tampering in some of these areas is well known. As an example, if the U.S. imposes high import tariffs (or quotas), there is little doubt that other nations will retaliate.

There is one viable option available to U.S. companies and organizations that does not have a down side, can never backfire, and surprisingly, often saves operating money in the long run. That option is to *improve the quality of goods and services in every way possible and as much as possible*. This book is about ISO 9000, a series of Quality Management System (QMS) standards that have been adopted by the European community and much of the industrialized world. ISO 9000 is a series of registration guidelines adopted and published by the International Organization for Standardization. The objectives of the guidelines are to promote worldwide standards to improve quality, operating productivity and efficiency, and to reduce costs. To be registered, a company must obtain the guidelines, adapt them to their own particular needs, prepare documentation, train employees, follow the documented procedures, and submit to an audit by an external registration organization. If the audit is successful, the company receives registration. ISO 9000 provides general guidelines unslanted to any particular type of business or industry. They are applicable to companies of 10 or 10,000 employees.

It is said that the United States (and much of the world) is rapidly becoming an information society and that decreased industrial output is simply natural evolution. It is even posited that the U.S. is in a post-industrial era. The authors disagree with this viewpoint, submitting that although many service sector jobs are very important, very few of them *produce wealth*. Wealth is produced by endeavors such as mining, agriculture, petroleum production, and manufacturing. Some of the most valuable scientific and technological developments have occurred in this country. Think about this list:

- transistors

- integrated circuits

- VCRs

- telephones

- television receivers

- microprocessors

- computers

- numerical control machines

- tape decks

Who supplies the manufactured versions of most of these U.S. inventions to U.S. consumers? It has been estimated that foreign nations have access to American research and development at costs ranging from one cent to five cents per dollar. So much for a bright future as an information society. Of course, the U.S. must continue to do basic and applied research and continue to develop new products and technology. That is not in question here. However, it would be nice to see more U.S. inventions manufactured in America. It would be even nicer to see many of them exported to the rest of the world.

Although manufacturing is very important, it is a fact that about 75% of American jobs are now in the service sector. It is also a fact that computer software is big business today. Because quality is a vital concern for all products and services, this book has been structured to support all three: quality systems in manufacturing, software, and services.

HISTORY

After World War II, the U.S. realized that Western Europe would have to rebuild, in an organized way, if long term stability was to be realized. The Marshall Plan was the first initiative which provided money for economic development from 1948 to 1952. Sixteen European countries formed the Organization for European Economic Cooperation to coordinate Marshall Plan aid. In 1961, this became the OECD (Organization for Economic Cooperation and Development). Another significant cooperative effort between the U.S. and western Europe was the North Atlantic Treaty Organization (NATO) which was established shortly after World War II in 1949. Although its major purpose was military cooperation and mutual defense, NATO also dealt with economic cooperation.

The European Coal and Steel Community (ECSC) was formed in 1952 and marked the beginning of market consolidation in Europe. The original group of six ECSC countries pooled resources and harmonized industrial policies. They formed a single economic market which was called the European Economic Community (EEC), also referred to as the common market. The ECSC became the EC (European Community) in 1958. The EC may also be referred to as the EU (European Union). Britain, Ireland, and Denmark joined in 1973, Greece in 1981, and then Spain and Portugal in 1986.

The EC picked up considerable momentum in 1985 when it proposed to accomplish 279 objectives by the end of 1992. Approximately 240 of these have been approved by the EC council of ministers. The twelve members of the EC constitute a market of over 300 million people and approximately 25% of all U.S. exports go to EC nations. Other nations have shown an interest in joining the EC.

To make the EC work as intended, full cooperation is required among the member nations. This cooperation includes areas such as a common currency, elimination of tariffs and quotas, uniform standards, sharing of information and technology, cooperative research efforts, education, energy and environmental policies, and legal codes.

One can imagine the chaos that would result if individual nations used unique systems of measurement. This is why there is an international standard of measurements (called SI) based on the metric system. One can also imagine that quality standards and procedures should also be universal, which when adhered to, would greatly simplify business. Otherwise all consumers would have to spend time and money to determine if goods and services would meet their expectations and needs.

In 1959, the U.S. department of defense developed MIL-Q-9858, which was a system for quality management. It was revised in 1963 and then adopted by NATO in 1968 as its Allied Quality Assurance Publication 1 (AQAP-1). AQAP-1 was adopted by the United Kingdom Ministry of Defence in 1970. In 1979 the British Standards Institute established BS 5750, which addressed the organization and documentation necessary to produce consistent quality. The first ISO (International Organization for Standardization, with headquarters in Geneva, Switzerland) committee on quality was formed in 1980. In 1987, 91 member nations (including the United States) of ISO published a series of five quality assurance standards (ISO 9000, 9001, 9002, 9003, and 9004). These are collectively known as ISO 9000 and are based on the earlier BS 5750. BS 5750 and ISO 9000 were harmonized in 1987 as equivalent documents and adopted by the EC. American relationship to ISO exists through the American National Standards Institute, which is a member of the ISO body. The ISO 9000 series was adopted verbatim by the United States as the ANSI/ASQC Q90 series. ASQC is the American Society for Quality Control.

The second edition of the ISO 9000 series was published in 1994. The revision was not substantially different, but it did emphasize effectiveness rather than mere conformance and it also addressed some vocabulary issues.

Currently, ISO 9000 is gaining considerable momentum. The ISO standard is the only quality system with international recognition, and it appears to be headed toward gaining universal acceptance. The FDA (Food and Drug Administration) and the big three U.S. automobile manufacturers are adopting ISO. The U.S. Department of Defense is approaching endorsement and ISO has already been adopted by the Ministry of Defence in the United Kingdom, by the Soviet Union, and by NATO. The countries of Japan, Canada, and Mexico are also moving toward adoption. The American medical products device industry is considering ISO 9001 as a replacement for its current standard.

THE ISO STANDARD

ISO 9000 is *not* a certification of quality, but an international protocol for organizing and documenting processes and procedures used to establish a quality system. It is not meant to replace or displace programs such as TQM (Total Quality Management) or six sigma (Motorola's award-winning quality management program). Specific product standards are defined by individual companies and their customers, not by ISO. The ISO guidelines provide the *structure* for developing and maintaining a quality system for manufacturing and services. This structure includes procedures, documentation, controls, and employee training. The structure is general and can be applied to any industry, regardless of the product or the service offered. ISO 9000 certification means that a company has invested the time to organize a quality system, has prepared the documentation, uses the documentation as demonstrated by quality records, has trained its employees, and has passed the scrutiny of a team of third-party registrars.

The earliest initiatives in quality control focused on inspection after a product was manufactured. ISO 9000 addresses a wide range of activities including:

- management structure
- training
- purchasing
- design
- sales
- inspection
- documentation
- testing
- installation
- repairs
- calibration of instruments

TQM, or quality management in general, is a smart way of doing business. ISO 9000 can serve as a model to help an organization develop a quality system and establish their own quality procedures and documentation. All of the important processes in a business are a part of ISO 9000. Any company that develops and follows the ISO 9000 standards is almost guaranteed to have a consistent quality system that will improve with time.

ISO 9000 certification is an increasingly vital selling point. It informs potential customers that a company is organized to produce consistent quality. Any company that does business with other companies can either spend time and money auditing each and every supplier to insure that they meet all requirements or insist that they have ISO 9000 registration. It usually costs too much and takes too long to investigate all potential suppliers. Suppliers also benefit by not having to be subjected to the rigorous requirements of each and every customer. ISO 9000 registration will minimize those untimely customer audits that devour company resources. In addition to the EC, the EFTA (European Free Trade Association), with its seven member nations, has agreed to adopt most of the provisions of ISO. In effect, this brings to 18 the current total of European countries that will probably demand ISO 9000 certification from their suppliers.

The rest of this book will cover the ISO standards in detail, relate how they fit into a quality structure, and detail their implementation. However, it will be useful for the reader to consider now the basic content of the five basic standards.

ISO 9000 (ANSI Q9000)—Guidelines for Selection and Use. This section explains fundamental quality concepts. It defines key terms and provides guidance on selecting and using (tailoring) ISO 9001, 9002, and 9003. It is the road map to using the other parts of the series.

ISO 9001 (ANSI Q9001)—Model for Quality Assurance in Design/Development, Production, Installation, and Servicing. This is the most comprehensive standard (20 elements) in the series and covers all elements listed in ISO 9002 and 9003. In addition, it addresses design and development.

ISO 9002 (ANSI Q9002)—Model for Quality Assurance in Production, Installation and Servicing. This section contains 19 elements that address the prevention, detection, and correction of problems that occur during production, installation, and servicing.

ISO 9003 (ANSI Q9003)—Model for Quality Assurance in Final Inspection and Testing. This is the least comprehensive of the standards (16 elements), and is devoted to the detection and correction of problems that occur during final inspection and testing.

ISO 9004 (ANSI Q9004)—Guidelines for Quality Management and Quality System Elements. This is used to develop and implement a quality system and to determine the extent to which an element is applicable. It is cross-referenced to the other 9000 standards and can be used to guide both internal and external audits.

This book also discusses quality systems for computer software and for services. ISO covers these in two guideline documents:

ISO 9000-3—Guidelines for the application of ISO 9001 to the development, supply, and maintenance of software. This can be used by companies that develop and produce computer software. It shares some elements with ISO 9001 and provides additional information specific to the software industry. It is important to note that an organization *cannot* be registered under this guideline. While it is a useful guide, they would actually be registered under ISO 9001.

ISO 9004-2—Guidelines for services. This builds on the quality management principles covered in the ISO 9000 to ISO 9004 series. It seeks to help an organization organize and manage the quality of its services. A service organization *cannot* be registered under this guideline but could be registered under ISO 9003.

As of this writing, ISO was developing these additional guidelines:

- 9004-3—service material
- 9004-4—managing quality improvement
- 9004-5—quality plans
- 9004-6—configuration management

Organizations might also find these ISO documents useful:

- 8402—quality vocabulary
- 10011-1—auditing
- 10011-2—qualifications criteria for quality system auditors
- 10011-3—management of audit programs

ISO 9000 was originally intended to be advisory. However, it was determined that a registration procedure was a distinct business advantage. Obviously, third-party registration carries more weight with current and potential customers than does a self-pronounced claim of conformity with ISO 9000 standards. Today, it is viewed as a necessity for unimpeded trade within the new European Community. Tomorrow, it will be an absolute necessity for doing business in America, and almost everywhere else in the world.

Chapter Review Questions

1. Which country emerged from World War II as the leader in manufacturing?
2. What ultimate advantage did Germany and Japan realize by having to rebuild their industries following World War II?
3. What two factors influence consumers the most when they make buying decisions?
4. What is the danger of a country imposing high import tariffs in an attempt to achieve a balance of trade?
5. What does the acronym ISO stand for?
6. What types of companies and organizations can benefit from ISO 9000 registration?
7. How large does a corporation have to be before seeking ISO 9000 registration?
8. List some endeavors that produce wealth. Why do activities in the service sector redistribute wealth rather than produce it?
9. What did the Marshall Plan have to do with the OECD?
10. What do EC and EFTA stand for?
11. What are differences among the ISO 9000 series, the ANSI Q9000 series and BS 5750?
12. Which one of the ISO 9000 standards is the most comprehensive?
13. Can computer software companies be registered under ISO 9000-3?
14. Can service organizations be registered under ISO 9004-2?
15. When was the second edition of the ISO standards published?

2

BENEFITS OF ISO 9000 REGISTRATION

INTRODUCTION

"The reward is in the journey" is both a book title and a philosophical viewpoint concerning human experience. It is an appropriate quotation to introduce this chapter because the journey toward ISO 9000 registration is a substantial reward in itself. The achievement of registration produces very tangible benefits, but the real gains are found in the dynamics of what happens to a company as it interacts with subcontractors, with customers, and within itself. The company adopts new ways as it journeys toward the culminating moment: the audit. It is a proud moment, after a company has done its homework and rightfully seeks its well-earned recognition, but the real significance is what happens in preparation.

The Spring 1994 *Journal of Engineering Technology* contains the abstract from a Delphi panel study of members of the eight professional interest groups of the Society of Manufacturing Engineers (SME). After three rounds, the SME participants identified 139 competencies needed by manufacturing engineers for the year 2000. Of the 139 identified, the number one competency was "understanding the importance of quality—the importance of doing it right the first time." Interestingly, the number two competency was "work in a team environment that requires compromise for the 'good of the whole.'" ISO 9000 is a total quality structure and fosters team efforts.

BENEFITS TO THE CUSTOMER

In the total quality movement, quality is defined by the customer. ISO 9004 states that the primary concern of any company or organization is to offer quality products or services that satisfy the customers' expectations. "For the customer, there is a need for confidence in the ability of the company to deliver the desired quality as well as the consistent maintenance of that quality." Hopefully, the following list of customer complaints and concerns can be totally eliminated:

- health and safety risks
- the need for incoming inspection
- high rate of latent defects
- false or exaggerated marketing claims
- unavailable goods and services
- goods and services that do not satisfy
- high acquisition cost
- high rework and reinspection costs
- high operating cost
- poor reliability and high maintenance cost
- high cost of disposal
- loss of faith in the company

Figure 2-1 shows the diagram of a quality loop as illustrated in ISO 9004. This diagram shows how a quality system applies to and interacts with all phases of the quality cycle, from the identification of ideas to disposal after use. Continuousness is the salient feature of the loop. Determining customer needs and expectations is inherent in the marketing function, but customer information should impact all decisions and policies. Customer feedback and information flow should operate on a continuous basis:

[Element 7.3] "All information pertinent to the quality of a product or service should be analyzed, collated, interpreted, and communicated in accordance with defined procedures. Such information will help to determine the nature and extent of product or service problems in relation to customer experience and expectations. In addition, feedback information may provide clues to possible design changes as well as appropriate management action."

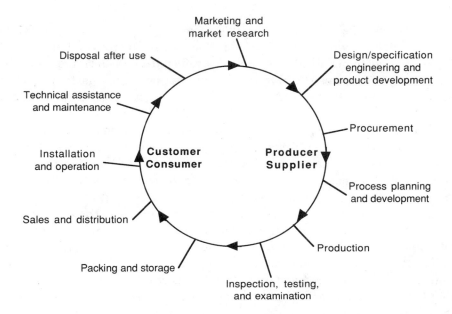

Figure 2.1 Quality loop (from ISO 9004)

[Element 8.9] "Periodic re-evaluation of product should be performed in order to insure that the design is still valid with respect to all specified requirements. This should include a review of customer needs and technical specifications in the light of field experiences, field performance surveys, or new technology and techniques."

[Element 16.3] "A feedback system regarding performance in use should exist to monitor the quality characteristics of the product throughout its life cycle. This system should be designed to analyze, as a continuing operation, the degree to which the product or service satisfies customer expectations on quality, including safety and reliability. Information on complaints, the occurrence and modes of failure, customer needs and expectations or any problem encountered in use should be made available for design review and corrective action in the supply and/or use of the item."

Customers benefit substantially by receiving an ever increasing value for their dollars when organizations place them firmly in their quality loops. This customer focus is sometimes referred to as "forming customer partnerships." A properly functioning supplier-customer partnership can yield some very good results. Who knows best what customers need and want?

Suppliers must learn to separate *feedback* information from *input* information (they are both very important). Some companies feel that an after-sale telephone call or mail survey is all the information they need from their customers. This feedback is important, but it is often inadequate. Customer input can positively influence design decisions and customer feedback after the sale can, at best, influence the next design cycle—which may be too late. Customer needs and priorities change and both input and feedback must be part of an on-going process. There is much to be learned about how customers use products. In some cases, focus groups of key customers have pointed out features that were not necessary. Significant simplifications resulted and the customers were delighted with the reduced cost.

A company should not forget about its internal customers. Any employee whose work follows another process should be regarded as a customer of that process. Such an employee, therefore, is one of the very best sources of feedback and input for improving that process.

In the final analysis, customers are the only authorities on company quality. Today, customers are better informed than they have ever been in the past, and well aware of their right to demand both quality and value. Quality of product and services is essential, but all areas of customer relationships are important. A happy customer is a repeat customer, one who will provide referrals and free advertising for the company that made them happy.

Some prime benchmarks are listed here to determine whether or not the customer is truly the focal point of any company or organization:

- A total quality climate exists. All contacts and dealings with the company reinforce a positive and professional attitude.

- All communications, including advertising, are based on facts. All dealings are ethical.

- All information given to the customer is clear and accurate.

- There is a strong commitment to solve customer problems.

- Customers are convinced that their feedback is important and that their business is highly valued by the company.

- Customers never feel the uneasiness of after sale abandonment.
- It is convenient and comfortable for customers to interact with the company before, during, and after the sale.
- Employees are empowered to satisfy customers in a timely fashion.
- Employees are courteous and knowledgeable.
- The quality of goods and services is world class.

Not one company in the entire world can legitimately claim that they meet every top benchmark. In fact, from time to time, there is slippage in most organizations in one category or another. The purpose of benchmarks is to help companies identify how they currently rate, to provide goals, and to serve as motivation for continuous improvement. When the benchmarks are used in an honest way, significant gains will be made and the partnership will enjoy the benefits.

BENEFITS TO THE SUBCONTRACTOR

In ISO 9000 parlance, the *supplier* is the company or organization using the standards to seek registration and their suppliers are called *subcontractors*. If Metafootware is a shoe manufacturer seeking ISO registration and it buys rubber heels from Laxxon, the following identifications are appropriate:

- Susan Murphy is a customer (she buys Metafootware shoes).
- Metafootware is the supplier.
- Laxxon is a subcontractor.

However, if Laxxon is also working towards its own ISO registration, then these identifications apply for their purposes:

- Metafootware is a customer.
- Laxxon is the supplier.
- Regal Rubber is a subcontractor (it sells raw materials to Laxxon).

Element 4.6.2 in ISO 9001 states that: "The supplier shall select subcontractors on the basis of their ability to meet sub-contract requirements, including quality requirements. The supplier shall establish and maintain records of acceptable sub-contractors." The data used as a part of the agreement with subcontractors and contained in purchasing documents include:

- type, class, style, grade, or other precise identification
- title or other positive identification
- specifications, drawings, process requirements, and inspection instructions
- requirements for approval or product qualification
- appropriate procedures, process equipment, and personnel
- the title, number, and issue of the applicable quality system standard

What are the benefits to subcontractors? For one thing, following the ISO guidelines will largely eliminate misunderstandings. Precise terms and specifications will allow subcontractors to fine tune their operations and eliminate nasty surprises, and their record as acceptable subcontractors will provide more stability for their operations. An increased level of communication and mutual understanding will benefit everyone.

Figure 2-2 shows two ways of doing business. In the "throw it over the wall" method, the subcontractor is isolated from the supplier and the supplier is isolated from the customer. The isolation is due to the invisible walls created by a general lack of understanding, feedback, input, and by vague agreements. What comes over the wall can be acceptable, barely acceptable, or unacceptable. It is as much a matter of luck as anything else. If it is unacceptable and bounces back over the wall, it is often accompanied by incredulity and ill feelings on one or both sides. In the partnership method, the subcontractor and supplier work as a team. They cannot have bitter disagreements because they work together for their mutual success. The positive aspects of the partnership between the supplier and customer were discussed previously in this chapter.

What happens when the subcontractor and the supplier are both ISO 9000 registrants? In this situation, the benefits are significant. Costly redundancies in auditing procedures can be eliminated. Visits between the two parties will deal with improved quality systems, further cost reductions, important innovations, improvements in productivity, and a strengthening of the partnership. These are all positive interactions and they advance the state of the art. It can be said that two teamed companies are smarter than either one. As an analogy on the negative side, consider a poor relationship between an architect and a contractor working together on a building. Because they are more adversaries than partners, much of their time together

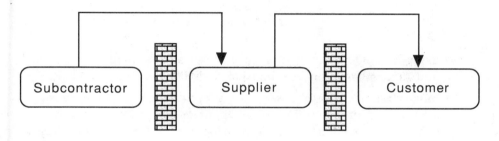

"Throw it over the wall" method of doing business

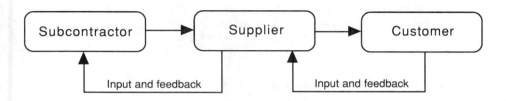

Partnership method of doing business

Figure 2.2 Subcontractors, suppliers, and customers

will be spent on complaints, annoying details, and the resolution of problems that should not have occurred. The architect and the contractor will be heavily engaged in telling each other how it *should* have been done and practicing finger pointing. Their building will never achieve its intended quality and value because it was condemned before it was built.

In the past, companies pressured their subcontractors by using a bidding process, and the subcontractors pressured their customers to order in larger volumes. The subcontractors also looked for ways to reduce their costs and were forced, in some cases, to reduce quality. They also passed cost pressures on to *their* subcontractors and a chain reaction of quality reduction sometimes occurred. Often subcontractors were dumped, without warning, after years of doing business with an important client. This produced a mindset that replaced loyalty and trust with cynicism and gray ethics. Today, partnerships that allow both parties to win are best accomplished by:

- eliminating the lowest bidder concept as the only criterion used in selecting subcontractors
- using ISO 9000 registration in place of audits, inspections, visits, and tactics based on intimidation and coercion
- having regular meetings with the supplier's team, including representation from manufacturing, engineering, process engineering, quality, and design in addition to personnel from the purchasing department
- using available technology, such as electronic interchange of information, to facilitate efficiency and accuracy
- participating in flexible arrangements which support methods such as just-in-time (covered in Chapter 6)
- seeking areas of mutual benefit where the parties can help each other solve problems, improve quality, and save money

BENEFITS TO THE SUPPLIER

As previously discussed, the supplier is the company or organization that has achieved, or is working toward, ISO 9000 registration. In one sense, it is not outlandish to state that the benefit to the supplier is that *the supplier can hope to stay in business.* This is a rather strong statement, but it does represent a future possibility. For example, the manufacturer of an electrical device hopes to sell his product through normal retail outlets but he has not gone through the process of UL (Underwriter's Laboratories) approval. What will happen when the marketing director approaches the retailers' buyers? Because of potential product liability problems, the buyers will not even consider handling the device. The manufacturer is dead in the water!

Currently, ISO 9000 registration is not required in the same way that the UL label is for certain products in the United States, but the world is moving in that direction. ISO registration will be required to do business in many parts of the world. The movement is gaining momentum and, as more countries demand compliance, even many small companies unfamiliar with ISO will be confronted with ISO registration.

Rather than adopting a "do or die" point of view, it will be useful to look at some benefits for suppliers who have, or will have in the near future, achieved ISO registration. Those benefits are included in the list below.

- ISO produces a marketing advantage. As general acceptance of the standard increases, this advantage will increase.

- ISO provides access to new markets. It is a world standard with wider recognition than the Malcom Baldridge Award, a U.S. award, and the Deming Prize, a Japanese award. (The latter two are awards whereas ISO registration demonstrates *conformance* to a worldwide standard. They are often compared, but it is important to understand the difference.)

- ISO can improve efficiency and reduce costs. This will vary according to the condition of the company when it begins implementing the standard. If a company was not very far along with any organized total quality system effort, then the improvements realized by implementing the ISO standards can be substantial.

- ISO will improve the quality and value of company products and services.

- ISO will increase customer satisfaction and repeat sales.

- ISO will enhance the company image.

ISO 9000 registration by an accredited registrar can be likened to the accreditation of colleges and universities. Because it is not always feasible for a potential customer (students and parents) to visit a school and investigate its standards, accreditation serves as a critical criterion when shopping for colleges or universities. Likewise, when businesses hire college graduates, they often accept accreditation as a guarantee of an acceptable level of standards. Often, people doing the hiring have negative feelings about an applicant with a degree from a non-accredited institution.

The major areas of the ISO guidelines include:

- management policy, responsibility, and review

- a documented quality system

- contract review

- design control, input, output, verification, and changes

- document control, approval, and changes

- purchasing, verification, and subcontractor assessment

- process control, inspection, testing, calibration, and records

- nonconforming product control, disposition, and corrective action

- handling, storage, packaging, and delivery
- internal audits
- quality records
- training
- servicing
- statistical techniques

This list demonstrates the comprehensive range of the ISO standards. Companies implementing systems of total quality, such as the ISO model, have all but eliminated the following leading causes for quality failures:

- no written instructions
- failure to follow written instructions
- unauthorized changes
- failure to remove obsolete documents
- lack of, or ineffective, corrective actions
- uncalibrated equipment

Additional benefits are more subtle but just as important. A prime example is the relative freedom from penalties associated with the loss of key personnel. The level of documentation suggested by ISO allows operations to continue in a smooth fashion when new workers must step into complex jobs. A tongue in cheek summary of the ISO guidelines is both humorous and appropriate:

- If it moves, train it.
- If doesn't move, calibrate it.
- If it's not written down, it didn't happen.

Chapter Review Questions

1. Who is the most important judge of quality?
2. What are some of the benefits of establishing supplier-customer partnerships?
3. Define the terms *customer, supplier,* and *subcontractor.*

4. Can one organization be a customer, a supplier, and a subcontractor? How?
5. Explain how ISO can simplify interactions between suppliers and sub-contractors.
6. Discuss how bidding can adversely affect quality.
7. Discuss some benefits of ISO registration for a supplier.
8. List the major areas of the ISO guidelines.
9. Explain how adequate documentation can help a supplier improve its operation.

3

ORGANIZING FOR ISO IMPLEMENTATION

INTRODUCTION

One of the most formidable tasks concerning ISO 9000 is getting started. Awareness of ISO 9000 can come in bits and pieces from many areas. The engineering manager might read about it in a trade journal article; the purchasing manager might hear about it at a trade show or conference; the vice president could have a conversation about it on the golf course. Frequently, awareness also comes from questions posed by customers to the marketing and sales division. Some examples of these questions are:

- What are you doing about ISO 9000?
- How long will it be before you are registered?
- Do you want to remain on our approved suppliers list?
- Did you know that your major competitor plans to be registered by the end of this year?

Awareness of the need to do something about the ISO standards leads to questions. There is considerable confusion in the areas of how the standards work, what an organization needs to do to get ready, and how much time and money will be needed. For example, a recent magazine article stated that once a company achieves ISO 9001 registration, it can choose to move on to ISO 9002 and ISO 9003 registration. This is *not* the way it works!

Actually, 9001 registration is the most comprehensive of the three, and one of the initial questions for many companies is: "Are we subject to the ISO 9001 or the 9002 standard?" The answer to this question depends on the organization's design function. A company with an engineering department could possibly be registered under ISO 9001. However, the function of the engineering department must be in the area of basic design undriven by specifications authored by the customer in order to be considered a true ISO 9001 application. If the engineering department works solely from customer specifications, then the application is ISO 9002.

If your company is in the software business, you will want to read Chapter 7 carefully. If you are in the service sector, then Chapter 8 is for you. Although ISO 9000 is usually associated with manufacturing, there is increasing activity for other types of organizations as well. If the current trends continue, ISO 9000 will become the most pervasive quality model of all time.

Another typical question is: "How long will it take to install an ISO 9000 quality system in my business?" This is a difficult question to answer because it depends on many factors. You can start with an estimate of two years and then subtract time if you already have the following systems in place:

- a comprehensive quality manual

- an internal audit system

- an approved supplier system

- a corrective action system (supplier, internal and customer)

- a regularly scheduled management review

- a comprehensive document control system

- an aggressive training program

- a hierarchical system of quality documents (quality policy, quality manual, area procedures, process work instructions, and quality records)

Companies uninitiated to the task often do a poor job of estimating the time and effort required to complete ISO registration. An organization can have a good quality manual, comprehensive procedures, and a strong quality department, but if these do not align with the ISO 9000 standard, there is still much work to be done. Companies already operating under military or governmental regulations are usually a large leg up on the process. This is because these regulations require documented procedures, training, auditing, corrective action systems and many other elements essential to the ISO 9000 standards.

Choosing the Coordinator

Most executives will respond to ISO awareness and related questions by delegating a manager to investigate procedures. Choosing this individual can be one of the more critical decisions made by upper management. This person is often at a fairly high level in management. It is absolutely essential that upper management understands that coordinating such a major effort is usually a full-time job. It is simply unrealistic to add this responsibility to an individual's workload without making judicious adjustments. It might be a part-time job in a very small organization, but in most organizations the task is large enough to be full-time.

What kind of an individual will make an effective ISO coordinator? This is a difficult question with no definite answer. However, some general guidelines include someone who:

- is interested in the assignment (perhaps a volunteer)

- understands quality and respects its value

- understands the company culture

- enjoys acting as a change agent

- has good interpersonal and communication skills

- works hard and has proven efficiency

- is respected by fellow workers

- truly believes in teamwork

The ISO coordinator can get started by attending an intensive training course that not only describes the ISO 9000 standards in detail, but teaches auditors how registration audits are done. There are a number of excellent lead auditor training courses that can fulfill that need. A local chapter of the American Society of Quality Control (ASQC) can provide a schedule of area courses. After the ISO coordinator is trained, he or she must develop the scope of the primary implementation effort. For a multi-site company, it is frequently advantageous to pick a lead site and bring the others along later. This keeps the implementation task small enough to be completed in a reasonable amount of time.

The employee who has been given the task of organizing the company for ISO registration will probably look for a seminar to attend, trade journal articles (with ISO 9000 in the title) to read, and begin investigating through professional networks. One of the best resources is an associate who has experienced the process and knows the pitfalls. Also, there are a number of good consultants who can provide guidance through the process, but care must be taken. Look for those who have experience at the actual implemention of an ISO 9000 system. There are many consultants who are trainers and can articulate well but who have no practical experience. Those who have been through the process are very valuable and can counsel the do's and don'ts of implementation. Other considerations are:

- What is the client success rate?

- How many successful clients can be claimed?

- Is an ongoing schedule of support after training available?

- Are written, detailed work instructions provided? (This is not a good practice. Process experts should do this, perhaps with the help of the consultant.)

- Do they have a broad breadth of experience, including the appropriate industry segment?

The next chapter provides more information and guidance for choosing and working with consultants.

The only way to get started is to *get started*. What is a reasonable amount of time to allow for the whole process? The answer depends on many factors, but the range of time for most operations is six months to two years, with the majority taking 12 to 18 months. The chief time factors are:

- management commitment and involvement

- availability of in-house resources

- training level of in-house resources

- funding for outside resources

- presence of a mature and comprehensive quality manual

- status of documented procedures

- existence of an approved supplier system

- existence of a documented design protocol (for ISO 9001)

- presence of an internal audit system

CHOOSING THE TEAM

The key to any major business initiative is the commitment and active involvement of upper management. ISO 9000 implementation and maintenance is no exception. Management must set the tone by becoming a visible force in the implementation process. Company-wide kickoff meetings are an excellent way for the entire organization to observe the commitment of management. Also, monthly or quarterly pep rallies can demonstrate that ISO 9000 is not simply another quality flavor of the month, so to speak. Companies have successfully used promotional tools such as an ISO 9000 employee of the month, ISO promotional mugs or T-shirts, ISO monthly newsletters, and so on. It is also effective to have the highest ranking member of management address the entire organization in quarterly updates to maintain the visible support of management. Many executives have tied management bonuses to the successful implementation of ISO 9000. There is nothing like money to breed enthusiasm and help drive the process. Beyond early planning, one key role of management is management review. This is covered extensively in another chapter.

Today, many organizations have made substantial reductions in their staffing to reduce costs. Some have gone overboard with this "lean and mean" approach to staffing, so there are few resources left to manage the difficult process of ISO 9000 implementation. The only way to respond to this situation is to put ISO 9000 high on the priority list and redeploy people from their current jobs to the implementation of ISO. To oversee the activities, there should be enough people in the organization to form an ISO 9000 core team. This core team should not spend the majority of its time on ISO, but it must commit the time for regularly scheduled meetings. It is ideal for the chief executive to be on the core team. His or her presence sends the clearest message about the importance of the effort.

Choosing an effective team is easier when the process of organizing for ISO is clearly understood. Three things in the real estate business determine all sales:

1. location

2. location

3. location

In parallel, ISO is based on three things:

1. documentation

2. documentation

3. documentation

The best stories are often those based in truth but use exaggeration to make a point and to add interest. With ISO, it is actually difficult to exaggerate the importance of the documentation process.

The quality manual is the master document for an ISO 9000 quality system. Those companies having an up-to-date quality manual (a living document) aptly describing their quality system have a leg up on most companies who have begun the process of ISO 9000 registration. Even if the company's manual does not address some of the elements of the ISO 9000 standard, it is better than starting from scratch. Developing a new quality manual from scratch can take up to six months especially for an employee who is not an experienced writer or who must learn about quality technology. This is an area where an experienced consultant can provide a substantial service. The quality manuals of most ISO registered companies share common system elements, and an effective manual can be developed by a consultant working with internal process experts in a reasonable period of concentrated effort.

If the process procedures affecting quality have already been adequately documented, then the implementation of ISO 9000 will be faster and easier. The degree of detail should be appropriate to the level of training or education of the targeted audience. It is not necessary to have greatly detailed, written procedures for highly skilled or educated personnel. Simple diagrams will suffice in many cases. However, if there is a significant turnover rate in the workforce, or if contract or temporary employees are frequently used, then the written procedures must be more detailed. Preparing these procedures can be time-consuming and this should be anticipated when developing the implementation time-line.

Much of the ISO 9000 implementation effort will be in preparing procedural documentation where none exists. This documentation should come from the process experts who are actually doing the work. These process experts might have limited writing skills and may need help to effectively transfer their knowledge to paper. Another consideration is that this large undertaking will require the use of project teams. If the people responsible for the implementation have not been trained in team dynamics, a training effort in this area will be needed.

Another important area for training is in the ISO 9000 standard itself. Everyone involved in the implementation effort must be aware of what the standard requires in his or her area of responsibility. In-house training personnel or professional ISO trainers can make employees aware of ISO standard requirements.

A typical question asked following the ISO standards training is: "Why is all this documentation necessary, and, how can we possibly write everything down that we do?" Happily, it is not necessary to write procedures in excruciating detail. Often a process flow chart with a small amount of text to describe each major step is enough. Process flow charts are very powerful tools that not only act as procedures, but reveal all of the non-value added steps that can choke an organization. This is just one example of how ISO 9000 supports the best practices of Total Quality Management. Documentation is meant to retrieve company knowledge stored only in the heads of the employees, on paper, or in a computer so that it is not lost in the event of retirements or major reorganization. It is also meant to reduce the worker-to-worker variation common in most organizations. You can apply a common-sense test to determine when the documentation is adequate. For example, ask: "If Sue retires tomorrow, will the written procedures for her job allow a new employee to step right in and fill her shoes?"

There needs to be a balance between documentation and training. For those organizations employing a few highly skilled or trained individuals, a great deal of documentation to define hour to hour and day to day activities is not needed. However, for organizations employing a large number of labor-intensive employees with a high turnover rate, documentation must be more comprehensive. The amount of documentation is often determined by internal audits. If auditors find significant variation from worker to worker, then the documentation is not as comprehensive as it should be. Good written procedures are very important, but without careful control they will not provide the desired results. Document control is an area that raises many questions. This issue is an important one and is covered in detail in Chapter 5.

Developing a system for controlling the purchasing of products from approved suppliers is another activity that can take the ISO team a long period of time. Not only must there be an approved supplier base, but all suppliers whose products or services impact quality must be regularly evaluated. Not all suppliers must be on the approved suppliers list, but instead only those whose offering can negatively impact quality. Metrics for evaluating suppliers must be developed, including:

- the number of problems found in shipments/number of shipments

- the number of parts accepted/number of parts inspected

- the number of latent defects (those found in production)/number of parts received
- the score on a supplier audit
- the score on a self audit

Whatever the metrics, records of the supplier scores must be maintained and procedures for removing suppliers from the list must be developed.

FINANCING THE EFFORT

As any other worthwhile endeavor, ISO implementation will require some funding. One of the first questions asked is: "How much will it cost?" That is another tough question since the answer depends on many factors. Top management must be prepared to pay for the training, redeployment of internal resources, and the hiring of outside consultants. The more money spent up front for these resources, the quicker the job will be done. This investment should be viewed as any other business investment. What is the payback? For example, if your customers are threatening to pull their business if you are not registered by the end of the year, then the money spent might keep you in business. If you are doing this because you know it is a wise course of action that will preempt external pressure about quality issues, then the monetary outlay can be considerably less. It all depends on your individual situation. Basically, the longer you have to complete your preliminary work, the lower the cost will be since you will be able to develop much of the needed expertise and achieve most of the development using existing personnel. On the other hand, if you are pressured for time and manpower, you will probably require more outside assistance.

The timeline can be shortened if you are in a position to reassign strong performers to the ISO implementation effort. Funds for training will also affect the timeline and an expenditure for experienced consultants can shortcut the process even more. ISO registration is rarely accomplished in less than six months. The most typical time frame for implementation is 12 to 18 months.

CHANGING PRIORITIES

One of the dangers to the successful completion of ISO implementation is distraction. Many companies have started reasonable and serious

ISO programs but later relaxed their efforts as other needs and ideas came along. Like the New Year's resolution, the effort in February is considerably less than it was in January. When March comes along, the resolution is forgotten. This is why top management must be committed to the process. It is their constant and visible presence that keeps the effort alive and produces reasonable progress.

Priorities change, crises come and go, fires have to be put out, and personnel come and go. Those employees involved with ISO registration must realize from the beginning that there are going to be distractions and that enthusiasm tends to wane. Chapter 9 covers how the Gantt chart can be used to produce a visible tool documenting the progress.

Chapter Review Questions

1. Which of the ISO standards is the most comprehensive?
2. Why does the achievement of ISO registration take some companies longer, even when comparing similar types and sizes of companies?
3. List some of the internal systems that should already be in place if a company hopes to achieve registration in a reasonable time frame.
4. Enumerate some desirable personal attributes for an ISO coordinator.
5. Give one example of how a multi-site organization could approach the ISO process.
6. What are some of the qualifications to look for in an ISO consultant?
7. Does an organization's ISO core team generally have to devote full time to the ISO efforts? Does the ISO coordinator devote full time as well?
8. What does a quality manual have to do with the ISO activities?
9. As far as ISO documentation is concerned, when are detailed written procedures unnecessary? When are they necessary?
10. Can written procedures reduce variations in quality? How?
11. How can internal audits determine the adequacy of existing documentation?
12. Why is the commitment of top management employees critical to the realization of the goal of ISO registration?

4

CONSULTANTS AND REGISTRARS

INTRODUCTION

The ISO process usually involves a company with outside consultants and registrars (auditors). An organization should use great care in selecting these advisors. The consultants can assist with the process of organization, documentation and meeting the ISO quality standards. Then, independent auditors can determine if the quality system is in compliance with a specific standard such as ISO 9001. If the audit provides evidence of compliance, the organization is registered under that standard. Self-declaration is another option. Here, a company adopts a quality system and declares itself to be in conformance with one of the ISO standards but does not submit to an external audit by a third-party registrar. This is not a popular option because it does not produce as many benefits attributed to registration. ISO registration will often open new markets, satisfy customer demands, and reduce the number of customer audits.

CONSULTANTS

The cost of hiring outside consultants is sometimes an area of concern, but the "muddle through" method can take longer and cost more. Perhaps the premier story about consultants is about a retired engineer who was called back by his former employer to look at an aggravating problem perplexing plant personnel. The engineer arrived at the plant and queried the supervisor for five minutes concerning the problem. He then requested a rubber mallet, proceeded to a cabinet containing some control circuits, and firmly whacked the cabinet with the mallet. The problem disappeared and production resumed. Later, the plant manager expressed concern over the consulting bill which had been submitted in the amount of $5,000. He telephoned the engineer and said: "Charley, I can't figure out your bill. I understand that all you did was whack a controller with a rubber mallet." The engineer replied: "Oh, I'm sorry. I should have itemized that bill. The charge for whacking the controller with the mallet is $100 and the charge for knowing that it needed a whack is $4900." In retrospect, the plant manager realized that the problem had already cost far more than $5,000 due to time wasted while muddling through it on their own.

The forgoing story is humorous but obviously most outside consultants should do more than whack a cabinet full of circuitry. If they are good, the consultants will save time, money, and prevent failures. Failures need to be avoided because they tend to dampen spirits and jade employees. In other words, a bad start leading to failure can have a negative effect, making the attempt worse than having done nothing at all.

Outside consultants often bring fresh ideas to their customers because their experiences have given them a rich and fertile base from which to draw. They can serve as the architects of change and provide the new structure and ideas needed to move a company ahead. As intelligent practitioners, outside consultants try not to repeat mistakes and false starts. They tend to be much more objective than insiders and make an excellent sounding board for various ideas expressed internally. On the other hand, outsiders are not familiar with the specifics of one company and its special culture. They do not experience the day to day changes initiated by their influence. Every company is unique, and good consultants understand this fact and tailor strategies to maximize the gains and minimize the pains.

Selecting a consultant or a consulting firm is a critical process that must be done carefully. It is important to see a consultant's client list to determine how much experience they have had with ISO 9000 and with similar companies. Make contact with some of their more recent clients and keep your perspective and sense of balance. It is not desirable to have too much of a good thing. Some consultants encourage "ISO overkill" which results in a cumbersome system that adds cost, is overly bureaucratic, and can have a negative impact on the cost and quality of product and services.

One way to get started in the process of choosing consultants is to send key managers and employees to ISO seminars and workshops since these are often conducted by firms that also do location consulting. If a good match and the required depth of expertise is evident, the next step is to investigate further and negotiate for the consulting services at your company site.

The customer-client relationship should be a comfortable one. Consultants should be good listeners and they should be flexible but firm when the situation warrants it. Enthusiasm and positive attitudes about the company and things in general should be sensed, along with dedication to clients, the industry, and the country. Ethics and basic values must align with both the company and the consultants. If prospective consultants insist that they are in possession of all of the answers, caution should be exercised. Be wary of comments such as:

- "Yeah, I know, firm XXX had the same problem and the idiots wouldn't listen to me so it cost them ten million dollars."

- "Well, the CEO over at ZZZ dumped this whole mess on the director of QC and the poor guy had a stroke and had to take an early retirement."

- "Listen, it's lucky for you that I have the time now to sign you as a client because I won't be able to do so next month."

- "Don't worry about a thing. I have done this a million times. I will have you ISO registered in six months."

- "There is no way that you can pull this off on your own. If you don't use my services you will have to find someone else."

Feeling comfortable with an outside consultant can be called "good chemistry." However, simple charm is not enough. Consultants *must* be professionally sound. Take time to investigate their experiences and credentials because some are not exactly what they appear to be on the surface. For example, a few industrial consultants have acquired a Ph.D. by paying fees to a so-called external institution, taking a few tests, and writing a thesis. A visit to some of the external degree granting institutions might reveal that the entire campus is in a suite of three offices. An inquiry into accreditation might reveal that the institutions are not accredited because "they prefer not to rely on federal and state funding." This is in no way meant to be a denigration of the vast majority of ISO and industrial consultants who are properly credentialed and potentially valuable, nor is it a general condemnation of external degrees, because some are granted by accredited and respected colleges and universities.

A consultant's knowledge of technology and processes is a plus, but it is not the determining factor. His or her ability to communicate with all levels of personnel is more important. Do not lose sight of what the consultant is being hired to do. If external expertise is needed to improve technology, hire an expert in that area. It is unrealistic to expect one person or even a specialized team to solve all company problems.

Employing a consulting firm with actual hands-on ISO 9000 quality system implementation experience is critical. Scholars and theoreticians can be quite glib, but there must be a foundation of pragmatic knowledge that generally comes from experience. In all ISO 9000 projects, there comes a time for difficult decisions and there comes a time to say "enough." Experience provides the wherewithal to make such decisions in a judicious fashion.

Most consultants prefer to work with an organization on a part-time basis because they will be engaged concurrently with other clients. This should be perfectly acceptable in most cases because a company may not necessarily need undivided attention for weeks or months at a time. A consultant's experience with other companies is part of the paid package, and as far as the actual consulting time for a typical ISO 9000 implementation project is concerned, an estimate of one person for one month is average for a company. This estimate is a *very low tolerance figure,* however, because it is dependent upon many company factors such as:

- the complexity of quality needs and systems
- the complexity of each company, its vendors and customers
- the complexity and diversity of products and services
- the current status of the quality system and its documentation
- special needs in areas such as training
- the capacity to respond to the consultant's requests in a timely fashion
- the skill and experience of the consultant
- the level of commitment
- the level of employee motivation
- the ability and desire to do some or most of the work with internal personnel

If the process is seen internally as consultant driven, problems will arise. The lion's share of talent and the capacity for ISO registration must be inherent in the company. Inside consultants are very valuable. In larger companies, there could be considerable internal expertise and it should be used to its best advantage. In any case, it is always best to develop a team combining inside and outside consultants.

No consultant, no matter how gifted and motivated, can compensate for fundamental flaws such as an adversarial relationship between management and labor. Flaws of this type are critical in the sense that they will offset and undermine any gains made. For example, there is the case of the company that received the Malcom Baldridge National Quality Award one year and then filed for bankruptcy in the next year.

After preliminary communications and the first site visit, the company has every right to ask about the consultant's initial views and his or her major expectations of what can and should happen. Consultants should be questioned as to how much time is needed to learn about the organization before recommendations are made. If the preassessment seems unreasonable, the company or the consultant or both have not yet reached a level of understanding and communication required for the partnership to be successful.

REGISTRARS

There are three types of evaluation for an organization's quality system:

- internal audits or self-assessment (first party)
- customer audits (second party)
- registrar audits (third party)

Many organizations use all three types. Some organizations have too much experience with the second type. For example, some companies supplying the big-three auto makers have complained that it is common to have several second-party audits per week. A successful third party audit providing ISO registration can significantly reduce the number of second party audits.

Third party registrar audits are typically the most comprehensive and credible to the rest of the world. However, any credibility is attributable to the registrar. The question is: "Who recognizes the registrar?" Registrar recognition is a matter of accreditation. In the United Kingdom the National Accreditation Council for Certification Bodies (NACCB) accredits registrars, while in The Netherlands this is accomplished via Raad voor de Certificate (RvC). The NACCB and RvC are government sanctioned by their respective countries. As of this writing, there is no United States government sanctioned accreditation agency.

Government sanction can be an advantage for conducting international business, but this is traditionally not the way the United States operates. In recognition of this problem, the American Society for Quality Control (ASQC) formed a separately incorporated Registrar Accreditation

Board (RAB) in 1989. In 1991, the American National Standards Institute (ANSI) incorporated RAB into their registrar accreditation program. The joint program resulting from the ANSI-RAB merger, although lacking official government sanction, is the most visible and recognized U.S. accreditation agency. Work is under way to use the National Institute of Standards and Technology (NIST), which is a U.S. government agency, to define and possibly certify the process of accreditation.

Ideally, one registration should enable a company to do business anywhere in the world, but unfortunately, it is not an ideal world. Universal registration demands agreements and assurances that all registrars' systems are equivalent. Such agreements are difficult to work out due to interference from national pride, politics, and memories of former disputes. Consequently, international companies can be faced with multiple registrations. Fortunately, there is some relief on the way. ANSI-RAB has a memorandum of understanding with RvC, NACCB, and JAS-ANZ (the Australia-New Zealand accreditation body). Also, an International Accreditation Forum has been formed with the promise of expanded international cooperation and joint agreements in the future.

Although it is not the purpose of this book to detail the accreditation process, it is prudent to cover some background information so that a supplier can make the best selection of a third party registrar. The following discussion assumes RAB accreditation. The process begins with the registrar filing an application. If the basic criteria are met as evidenced by the application, RAB conducts an audit at the applicant's facilities and performs an observation of an actual supplier audit as conducted by the applicant. The RAB team files a report that includes their recommendation for granting or withholding accreditation. Any corrective actions agreed to during or after the audit are evaluated by the team. The RAB accreditation council then reviews the report, the corrective actions, and other relevant information, and if accreditation is granted, a certificate is issued to the registrar who may then advertise the achievement. Accreditations are valid for four years, after which a complete reassessment is required. Surveillance audits are conducted at one-year intervals.

RAB uses the same criteria as those used by countries in the European community and in the European Free Trade Association:

- ISO Guide 40 (general requirements for the acceptance of certification bodies)

- ISO Guide 48 (guidelines for third-party assessment and registration of a supplier's quality system)

- EN 45012 (general criteria for certification bodies operating quality system certification)
- ISO 10011 (guidelines for auditing quality systems)

Figure 4-1 shows a schematic of the accreditation and registration process.

RAB requires that registrars use certified auditors. ISO 10011 discusses the qualifications for auditors. The basic requirements include:

- general education
- training in auditing quality systems
- training in the ISO 9000 standards
- work experience in the areas where the auditor will function

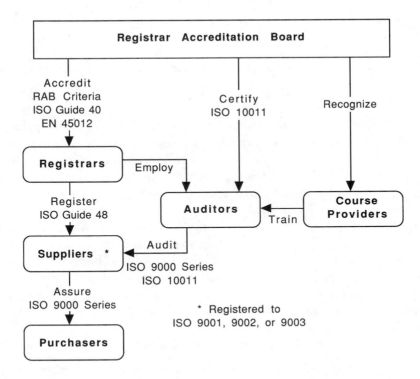

Figure 4.1 Accreditation and registration process

The grades of RAB certified auditors are:

- QS-PA (quality systems provisional auditor — entry level)
- QS-A (quality systems auditor — experienced level)
- QS-LA (quality systems lead auditor — qualified to lead audit teams)

This certification program recognizes various kinds of audit experiences. The following list is just a few of the standards RAB considers equivalent to ISO 9001 or 9002 for the purpose of granting audit experience credits only:

- ANSI/ASME N45.2
- ANSI/ASQC C1-1985
- Boeing D1-9000 (Rev. New 1991)
- DEC STD 017-1 (17/06/91)
- Ford Q101
- MIL-Q-9858A
- Motorola Corporate Quality System (March, 1991)
- Snap-On Tools Supplier Quality System (Rev. C)

In other words, auditing to these standards can be credited as auditing experience toward meeting the requirements for auditor registration. RAB certified auditors must maintain and expand their skills through experience and additional training. Experiences are documented in a log that must be submitted to RAB each year along with the auditor's application for continued certification.

Registrars typically become long-term partners, so it is critical to make the best possible selection. Registrars are required to perform periodic surveillance (usually every six months) to determine whether or not a supplier's quality system is being maintained. They usually require a complete re-audit after some period of time, usually three to four years. Choosing a registrar should be based on:

- the demands of your current customers
- perceived future needs including new products, additional services, and expanded markets

- the registrar's accreditation(s)

- the registrar's track record and stability

- the closeness of fit between your organization and the registrar's special skills and experiences

A registered supplier can use its registration, including the symbols of the registrar and the accrediting body, in its publicity, in its advertising, and on its letterheads. However, it cannot use registration statements or symbols on its products or in any way that could imply product quality. This is because registration applies to the supplier's quality system but not to any individual product or service.

In some cases, a registrar can provide a company with additional certification. Underwriters Laboratories (UL) is a noteworthy example. UL offers International Electrotechnical Commission's Quality (IECQ) assessment and National Electronic Components Quality (NECQ) assessment for electronic components. UL auditors can evaluate a supplier's quality system according to ISO 9000 requirements and, optionally, a supplier's electronic components can be certified to national or international specifications. IECQ certification means that components can be accepted internationally, often without the need for further inspections. NECQ certification provides recognition in various U.S. industries including original equipment manufacturers of commercial and military products. IECQ/NECQ certifications can save time and costs by eliminating duplicate assessments.

UL is perhaps best known for its label on electrical products. This is a separate function and it must be clearly understood that UL ISO 9000 registration does *not* indicate UL listing, labeling, recognition, classification or any other product certification except as noted in the prior paragraph.

UL has been granted RvC accreditation as an ISO 9000 registrar. It has also established memoranda of understanding with other registrars:

- American Gas Association Laboratories (AGAL)

- British Standards Institution (BSI)

- Bureau of Commodity Inspection and Quarantine (BCIQ—Taiwan)

- Canadian General Standards Board (CGSB)

- Factory Mutual Research Corp. (FM)

- Instituto Nacional de Metrologia, Normalizacao E Qualidade Industrial (INMETRO—Brazil)
- Japan Quality Assurance Organization (JQA)
- NSF International
- N.V. KEMA (The Netherlands)
- Quality Management Institute (QMI—Canada)
- Singapore Institute of Standards and Industrial Research (SISIR)
- Standards and Industrial Research Institute of Malaysia (SIRIM)
- Standards Australia (SA)
- Standards Institution of Israel (SII)
- Underwriters Laboratories of Canada (ULC)

Because of its relationships with other organizations, UL can assist a supplier to obtain multiple registration. This is advantageous when customers are located in more than one country and/or when market expansion is desired.

Chapter Review Questions

1. Why do few companies choose self-declaration for ISO compliance?
2. List some of the things that you should look for in an outside ISO consultant.
3. What is ISO overkill?
4. Is it mandatory for outside ISO consultants to work full-time with their clients throughout the entire ISO effort? Why?
5. Can a good ISO consultant take care of the entire ISO process for his or her client? Why?
6. Why is it a good idea to ask for an initial report after the first visit by an outside consultant?
7. What is the difference between an ISO consultant and an ISO registrar? Can a company hire one person or a firm to serve in both capacities?
8. Explain the differences among first, second, and third party audits.
9. Does a United States government agency currently exist that accredits ISO registrars?
10. Describe the relationship between ASQC and RAB.
11. What is the difference between a surveillance audit and a reaudit. What are the typical time intervals for each?

12. Discuss some of the important considerations for choosing an ISO registrar.
13. How can a company use the fact of its ISO registration? How can it not?

5

THE ISO 9001 ELEMENTS

INTRODUCTION

ISO standards 9000 and 9004 provide quality management guidance for all organizations. 9001 through 9003 are used for external assurance in contractual situations, but 9001 is the most comprehensive and includes design, development, production, installation, and service. A company's quality system should only be as comprehensive as required to meet their quality objectives. There are six factors to consider when selecting a model:

- the design process and its complexity (if yet to be designed)

- design maturity (as proven by testing or experience)

- the production process and its complexity (new or proven)

- the characteristics of the product or service (identify critical issues)

- safety (potential failures and their consequences)

- economics (the five preceding factors balanced against costs)

This chapter has been carefully organized for maximum readability by grouping the topics and concepts logically. This method made it impossible to order the element numbers consistently and arithmetically. For example, 4.18 is covered before 4.3.

THE ROLE OF LEADERSHIP

4.1 MANAGEMENT RESPONSIBILITY

The involvement of upper management is crucial for every quality initiative, project, or program. Without it, a quality culture change, such as embracing ISO 9000 as a new business system, is doomed to failure. ISO 9000 implementation can be delegated to responsible project managers by upper management who must continue to show support through a visible presence (read as *participation*).

Quality Policy

Management's first responsibility in their commitment to the ISO 9000 quality system is to define the organization's quality policy. All other quality documents are derived from this document, setting the tone for the organization's business conduct (see the next section of this chapter). The policy needs to include a statement of management's unwavering commitment to quality, to satisfy the customer's needs, and be written so that all in the organization can relate to it. It must also be suitable for the company's business without resembling a generic "motherhood and apple pie" statement.

The quality policy should also include quality objectives that the organization must strive to meet. These may be stated in terms of meeting customer needs, reducing costs, meeting competitive challenges, and so on. A generic quality policy is stated below:

"It is the policy of Company XYZ to supply its customers with defect-free materials. In order to achieve this result, we will use continuous improvement techniques to improve all processes in the company. Our aim is to become the top supplier of widgets in the global marketplace. We will use the ISO 9001 quality system to insure that our processes adhere to the best systems in the world. We are committed to this policy and require all employees of XYZ to comply with this policy."

Better defined objectives for quality may be listed in a separate document, but each organization's objectives must cascade down from the company quality policy. High-level quality objectives can be as follows:

• reduce customer complaints

• reduce inspection costs

• improve customer satisfaction index

• reduce scrap costs

- capture an increasing share of the "XYZ" domestic market

- reduce inventory

Management also has to insure that the quality policy is understood, implemented, and maintained at all levels of the organization. This can be done by posting copies (signed and dated by the chief executive of the organization) in highly visible areas, by reviewing it regularly in group meetings, or by printing it in documents that are regularly reviewed by the entire organization. Whatever method is used, everyone must be able to articulate the concepts of the policy as well as point to an actual approved copy of the document.

Organization

Management is responsible for defining and documenting the quality responsibility, authority, and interrelationships of all people in the organization, including the chief executive officer as well as the operators of equipment. This can be done with organizational charts and/or by defining job descriptions for each position. The job descriptions can be for main job groupings such as:

1. production operator

2. quality inspector

3. design engineer

4. engineering technician

5. scheduler

6. marketing manager

A typical job description is shown below:

It is the responsibility of the *technical manager* to assign product lines for each production engineer and technician. The technical manager must insure through quarterly reviews that the engineers and technicians are improving the quality of the products and processes for which they are responsible. The technical manager with his staff must develop the metrics for evaluating the success of the continuous improvement efforts. These efforts should result in reduced costs and improved customer satisfaction as evidenced by the monthly cost statements and customer surveys. The technical manager must work closely with the quality manager, marketing manager, and design manager to insure that customers are being properly serviced and that no customer concern goes unanswered. The technical manager receives his direction from the plant manager who must respond to the ever-changing needs of the business.

Resources

Management is responsible for identifying resource requirements to insure adequate support of the quality system, including: process equipment, measurement devices, facilities with the proper environmental controls, resources for proper documentation and quality record storage, and training for all personnel. A specific requirement defined in this "Resources" section is to provide for internal quality audits. When management is audited for *management responsibility* they will be questioned about their role in identifying the requirements of the organization and supplying all of the proper resources to meet these needs.

Management Representative

A management representative must be chosen to be responsible, regardless of all other duties, for the operation and maintenance of the quality system. This person is usually the highest ranking member of the quality organization (vice president of quality, quality director, quality manager, and so on), but is not necessarily in the quality organization. This person must be authorized to act independently insuring that everyone is following the guidelines required by the documented quality system. In small operations the owner of the company sometimes accepts this responsibility. In medium sized operations, where no separate quality organization exists, the manufacturing or technical manager may be appointed to fill this role. In larger organizations, it is usually the head of the quality organization as stated above. It is also the responsibility of the management representative to report on the performance of the quality system and to demonstrate how the quality system is being continuously improved.

Management Review

Management has the important responsibility of periodically reviewing the quality system to insure its continued suitability and effectiveness. This is a formal review and normally occurs no more than once per quarter. Some operations try to meet this obligation by discussing quality in weekly production meetings, but this is not the intent of this requirement. It is intended to be a time when quality system performance is the only topic and strategic planning at the highest levels of the organization can take place.

The meeting must be scheduled and chaired by the chief executive in order for the review to meet the intent of the standard. A typical agenda for a quality review is:

1. customer survey results

2. customer complaint analysis

3. quality improvement project status

4. quality cost performance

5. results of internal audits

6. upcoming training requirements

7. competitive analysis

4.2 QUALITY SYSTEM

This element requires a documented quality system that starts with a quality manual. The quality manual should be an accurate and complete description of the quality system and serve as a reference for day-to-day operational questions and as a guide to quality improvements. Some things that should be avoided when planning a quality manual include:

- attempting to borrow a quality manual from another organization

- assigning the development of the manual to one person

- making the manual more complicated and detailed than needed

- assuming that once it is done it will be cast in stone

Writing the quality manual, at first glance, can appear to be a formidable task, hence the temptation to borrow one. (Why reinvent the wheel? Because their wheel won't fit your cart!) For example: Let's give it to Joan. She is not busy right now. Unfortunately, Joan does not have more than a small fraction of the information needed to write the quality manual, and Joan does not establish level one policies in any case.

It is acceptable to look at other manuals to develop some feeling for what they contain and how they are structured, but it is not necessary. The way to begin is to commit the time and effort to develop the manual, pick a team and assign tasks, and develop a time line and set priorities. The first organizational meeting of the quality manual development team should set some concrete goals, but the team should be cautioned not to bog down in details. It's a good idea to make specific assignments to the team members to be completed before their second meeting. Some sample assignments are:

- Prepare a one-paragraph statement about the quality policy for your area.

- Gather all existing quality documents for your area. This would include quality records, organizational charts, flow charts, personnel charts, procedures, and so on.

- Make a list of those items that you feel should be added to document quality procedures in your area.

The second team meeting will be confusing if minute details are discussed. This meeting must be used to develop a skeleton view of the quality manual so people know that their specific concerns will be discussed after the basic structure is agreed upon (their issues will eventually become the "flesh on the skeleton"). Generally, quality manuals are composed of Level I documents and can be written to follow the ISO structure:

Level I — This is the overall company policy and states objectives for quality and customer focus. The overall company organization chart designates the major areas of responsibility. It contains references to lower levels, describes how management reviews will be performed, how the manual was prepared and will be revised, and describes the intent and scope of the total quality system for the company.

Level II — This section is specific to departments or locations within the company. Departmental organizational and flow charts show responsibility for quality functions, while department objectives should be stated and referenced to the Level I policies and objectives. Corrective action and quality enhancements are discussed. References should be made to the Level III documents, their method of control, and how they are revised. The Level II and Level III documents are usually not in the quality manual.

Level III — These documents are operation specific. Work instructions and procedures for inspection, handling, and calibration including the description and use of all appropriate Level IV forms are included. The work instructions are controlled documents that are available in the work areas where they apply and though they are generally not located in the main body of the quality manual, they are identified there.

Level IV — These include the checklists, forms, and records necessary to complete the Level III work instructions. The forms and checklists are controlled documents that are available in the work area. Once the forms or checklists are written upon, they become quality records and are subject to the requirements of Element 4.16.

Thus, the quality manual is usually made up of the Level I documents but is enabled through the other documents, written procedures and forms. The quality manual should also contain a revision page, an index, and a glossary of company terminology. Some companies structure the manual using an ISO standard as a guideline in an effort to make an external audit smoother and easier. Another idea is to use page one of the quality manual for documenting and dating all revisions to make re-audits easier. A quality manual is a form of a top-down strategic planning document. Ideally, it should never be started until top management has committed to an overall quality policy.

4.18 TRAINING

In Element 18.1.1, ISO 9004 states: "Consideration should be given to providing training for all levels of personnel within the organization. Particular attention should be given to the selection and training of recruited personnel and personnel transferred to new assignments." It is not necessary to provide training to experienced personnel when their experience and performance clearly demonstrate the skills and knowledge needed to perform their assigned tasks. In fact, telling people what they already know is at best a waste of time and can be a source of irritation. A documentation system should be in place that cites work responsibilities and operator experience. In some cases, worker qualification might be required as verified through periodic tests or inspections (this is common with functions such as welding).

Workers should be aware of their job descriptions because an external auditor may ask any worker if he or she knows about specific methodology and if required training has been provided. Production supervisors and workers must be aware of all the information that they need to perform their jobs:

- use of tools and instruments

- common causes of measurement errors

- sources of process variations

- use of machines

- the documentation for their area

- safety rules, equipment, and procedures

- quality factors for their area

- specific job skills (welding, print reading, soldering, and so on)

Training in basic statistical techniques should also be considered for these workers. The knowledge required by production personnel will vary from company to company; however it might be prudent to consider teaching them some basic topics such as sampling, mean, variance, standard deviation, and control charts.

Experience has shown that training is most effective when it is specific to job function and when it is broken into manageable chunks. The attention span of most human beings is about 20 minutes, which makes one wonder about the efficacy of marathon sessions that last for three days or more. Obviously, there are occasions when intense training is mandatory; however, when possible, more frequent short sessions are usually better. Also, the sessions should be conducted in an informal manner with ample opportunity for audience feedback and questions.

How training is conducted varies widely from company to company: a large corporation might have a department of training with its own executive level manager and a staff; a small company might use the president, or the head of engineering, or coworkers as part-time trainers. However it is accomplished, it should be documented.

Training can also be conducted by persons outside of the company and can take place at outside locations. Training consultants and vendors are two possibilities. In the case of vendors, it is sometimes prudent for a company to negotiate training as a part of the conditions of sale for new equipment. Again, documentation of the training is important. Do not overlook the technical schools, community colleges, and universities when training needs are apparent because many of these can arrange special short courses and will teach them at the company location.

An effective arrangement with an outside trainer requires some careful planning. The specific training objectives must be established within the company before looking to the outside. Here is an example of what can happen. The telephone rings and the secretary answers:

"Hello, engineering technology department, may I help you?"

"Yes, I want to talk to someone about a drafting course."

"Please hold."

"Hello, this is Bob Striker. How can I help you?"

"Mr. Striker, this is Phil Stewart down at ABX Bolts. Are you familiar with our company?"

"Yes I am Phil. Please call me Bob. What can I do for you?"

"Bob, I have about 12 employees who really could use a drafting course. It's my understanding that your college teaches drafting and I was wondering if you could help us out?"

"Yes Phil, we do teach drafting, mostly CAD and descriptive geometry, and we probably can work something out."

"Hey, that's great. Let's get together, say sometime next week and. . . ."

As the story turned out, Phil and Bob had lunch together the following week. Luckily, Bob figured out that Phil's company did not need a drafting course. He was actually talking mostly about specific training in geometric tolerancing. Bob then admitted to Phil that another person in his department would be better suited for this type of training. Bob gave Phil her name and telephone number and the process started over.

Now that the specific area of need was identified, Phil managed to do a little homework. He spent some time with the appropriate departments and supervisors and was able to determine the specific skills that the employees needed, he was able to put together a profile of the employees who would attend the training, and, working with this information, the college instructor developed an outline of five one-hour sessions. She faxed the outline to Phil who shared it with the departments. Suggestions and changes were faxed back a few days later, and finally, a revised outline was approved and the sessions were conducted on the company premises. The instructor did a good job and came prepared with some effective transparencies and handouts. The employees rated the sessions highly and enjoyed the interaction that they had with the instructor.

Phil was a busy man. In the beginning, he hoped to avoid involvement with the details of this "training thing," because he assumed that the college knew what it was doing and that everything would work out correctly. In retrospect, Phil realized that it had cost him several hours to formulate the details but the results were well worth his efforts. He could imagine what probably would have happened if he had arranged for a standard drafting course and vowed to do his homework *first* the next time.

ISO 9004 also describes executive and management training. These personnel should understand the total quality system of the company and have knowledge of the quality tools and procedures in their areas of responsibility. They also should have the skills necessary to evaluate the effectiveness of the system.

Technical personnel are also mentioned and training needs are cited in areas beyond those normally considered to be the primary quality functions. Some additional areas include:

- marketing

- procurement

- process engineering

- product engineering

- statistical techniques such as process capability studies, sampling, data collection and analysis, problem identification and analysis

- corrective action

Elements 18.3.1 through 18.3.4 of ISO 9004 treat personnel motivation and personnel quality awareness. Workers should understand their tasks and realize that they are a vital part of the overall quality of what their company has to offer. Effective team training is necessary because a true customer focus can only be achieved when all employees feel that they are important team members. Quality improvement teams can unleash the creative potential of an entire organization with training that includes team dynamics and incorporates facilitators.

In a healthy company, there is a spirit that discourages negative attitudes and poor performance. Motivation efforts should be directed toward all segments of the workforce including management, marketing, design, documentation, manufacturing, inspection, shipping, and service. Motivation is nurtured by enthusiasm and dedication that begins at the top, predicated on a positive attitude and is best produced by example.

New employees should receive an indoctrination to the total quality system accompanied by an employee quality handbook that can be a subset of the formal quality manual. Generally, the employee handbook would not be a controlled document since it would be prohibitively expensive to keep all copies updated and accounted for. This handbook should emphasize how employees are empowered to make suggestions, including corrective actions, and how they fit into the overall structure. Periodic refresher programs for experienced employees can emphasize recent changes and the challenges facing the company today and tomorrow. Managers and supervisors must have the authority to reward and recognize their workers for outstanding contributions and for loyal, long-standing service.

CONTROL OF EXTERNAL INTERFACES

4.3 CONTRACT REVIEW

Contract review is a key element because it is the starting point for all business. The agreement between the customer and client must be well defined so that there will be no surprises when the offering is delivered. Anyone who has ever built a house can appreciate what a poor contract review can mean. If the number and location of electrical outlets, light switches, cable TV ports, appliance colors, carpeting selections, and so on are not specified by the client, then somebody must do it for them. A well thought out contract review can minimize problems down the line.

Cross Functional Contributions

Successful contract review can only be achieved if customer requirements and expectations are well defined and communicated to all who are impacted. Everyone who can contribute to customer satisfaction must be involved with the written agreement because the organization's capability to meet agreed upon specifications hinges on everyone's input. Marketing claims must be backed up by organizational capabilities. Any sales brochures, manuals, or other documents used by the marketing and sales organization should be controlled as well; however, they are not formally controlled documents since there is no practical way to retrieve all outdated copies.

What is more frustrating than falling in love with an item in a catalogue, ordering it, and then discovering that the product line has been discontinued and there are no satisfactory replacements? The sales brochures must be kept up to date with product enhancements and must not make false claims that can not be substantiated with functional capabilities.

Modifications to Contracts

Records of customer transactions must be maintained. Whenever a contract is modified in any way, it must be written down and retained, including items such as memos, faxes, and verbal or phone conversations. These can be in the form of a note to file with a signed and dated record of the conversation. If a formal contract is being used between the parties, the formal contract must be revised with the changes.

4.6 PURCHASING

Purchasing is one of the key elements of interface with the outside world. It is very difficult to deliver a quality product or service if the purchasing function is not effectively controlled. It is important to have well-defined procedures that describe the process by which supplies and services are ordered and delivered. These procedures must first define what information must be present on a purchase order, such as:

- an accurate description of the items

- the quantity desired

- the part numbers

- the revision levels of the drawing or specification

- the quality requirements

- delivery time

- delivery mode (truck, train, plane, mail, and so on)

- packaging requirements

Inaccurate information in any of these categories can cause quality troubles in later stages of production or service.

Purchase orders must be reviewed and approved before final processing. This standard also requires that suppliers be selected based on their ability to provide products and services that meet quality requirements. Records of supplier's performance must be established and maintained which means that a system must be developed to investigate potential suppliers, evaluate current suppliers, and remove problem suppliers from an approved suppliers list. This requirement of the standard usually causes a great deal of consternation to the organization, however, if it is properly handled, it can produce significant rewards:

- reduction in incoming inspection costs
- reduction in cost of administering raw material returns
- improvement in production scheduling
- reduced processing costs due to defective raw materials
- improved product first pass yields

The old saying, "you can't make a silk purse out of a sow's ear" applies in this case. If you are starting with a high percentage of substandard parts, your manufacturing costs are bound to escalate dramatically. How can you make a world class factory operate when you are constantly dealing with substandard raw materials? Many companies have implemented automation to improve their processes, but have not addressed the raw materials issue. This is process control suicide.

One issue concerning the use of robots is that they must have extremely consistent materials to work with because, without highly consistent feed stocks, robots will just make scrap much quicker than hand assembly. With hand assembly, the operator can intervene and make minor adjustments. With automation equipment the mindless robot does not care if 75 percent of the product it makes is useless. Usually, substandard raw materials in an automated process can result in having to modify the equipment to get any production at all, and this frequently results in higher production costs than those incurred with conventional hand assembly.

Consider the case of the company that had a huge opportunity with a large customer to make high-tech electrical components. They quoted a price to the customer under the assumption that automation would bring the processing price down far enough to make a good profit. Unfortunately, they did not address the raw material issues and discovered too late that the automation worked well with very consistent parts, but almost not at all with less than consistent parts. They struggled with the problem for three years before they finally abandoned the robots and went back to hand assembly. If they had dealt with the component variation problem from the beginning, they would have either insisted on reduced variation in the components, or not considered automation in their pricing structure. This lack of foresight resulted in a three million dollar loss on the project, ill will with the customer, a lost business opportunity, and numerous lay-offs for some very hard-working people.

4.7 CONTROL OF CUSTOMER SUPPLIED PRODUCT

The intent of this element is to insure that customer products sent to a vendor or subcontractor are not subjected to damage or deterioration. The contract defining the responsibilities of both parties should spell out the terms of the agreement in detail. Written procedures should be followed to check the product when it arrives and to notify the customer of any damage, deterioration, or other irregularities. There might be requirements for special handling, inspection, calibration, maintenance, storage, or handling. If the product must be stored in environmentally protected areas (i.e., refrigerators), then the associated environmental controllers and recorders must be included in the calibration system. Any problems encountered during processing should be documented and integrated into the corrective action system to realize effective remediation. The customer is ultimately responsible for the quality of any material that he supplies. In light of the ISO standards, there is nothing special about customer supplied product that permits circumventing any part of the total quality system and its procedures.

In this area, quality records are just as important as the records used for the materials coming from all other sources. Here again, a partnership type of arrangement with input and feedback will provide consistent quality and improvements over time.

4.19 SERVICING

"Where service is specified in the contract, the supplier shall establish and maintain procedures for performing and verifying that servicing meets the specified requirements." From a customer viewpoint, servicing can be the most important quality issue, but, unfortunately, it is often an afterthought. Although it is often a vital function, it might not receive as much attention as that given to other processes such as design and production.

The correct time to consider servicing is early in the design cycle. If the product will be serviced, can it be designed in such a way to make servicing faster, easier, and less costly? Just as design for manufacturability, design for serviceability simply makes good sense. This is why the modern design team consists of key representatives from all major parts of the organization.

If servicing will require special tools, equipment, or software then these materials must be verified, their design and functionality must be tested, and adequate documentation should be prepared to assist and guide those doing the servicing. All of the materials needed for servicing must be available in a timely fashion. The documentation package for servicing could include:

- installation instructions
- troubleshooting guides
- assembly and disassembly instructions
- operational parameters
- technical specifications and verification procedures
- parts and subassembly listings
- drawings and/or schematics
- operating guides for special test equipment and/or software
- all relevant safety information

Calibration and calibration records are required for the measuring and test equipment used in field servicing and servicing documents should be controlled as explained in a later section of this chapter. Other factors that should be considered include:

- preventive maintenance schedules and instructions
- service bulletins and product alerts
- telephone support
- electronic bulletin boards
- field upgrades and retrofits
- recall procedures (hopefully never needed)
- service training requirements

The logistics of servicing should be carefully planned and agreed upon with special consideration given to the possible overlap of responsibilities among the supplier, field representatives, outside servicing organizations, distributors, and users. The supplier should insure that all parties have the proper materials, documentation, and training needed for effective servicing, especially when servicing has a potential impact in the areas of safety and liability.

DESIGN CONTROL

The design process is the deciding factor when a supplier is considering using either ISO 9001 or 9002 as a model. The design process complexity and difficulty for new products or services must be determined, as well as the maturity of the design as demonstrated by field experience or performance testing. If designs are not already mature and proven, and if the design process is not trivial, then design control as required in Element 4.4 is appropriate and desirable. The primary deciding factor is whether or not the supplier receives product specifications from the customer. In the ISO 9000 realm, design refers to the supplier developing a new product based on internal specifications. A supplier can test market a product and modify the design specification according to customer feedback but the supplier is still in control of the design.

In order to conform to element 4.4, the supplier must verify designs to insure that all requirements are met. Design critically affects quality aspects such as performance, safety, reliability, suitability, and dependability, and can apply to hardware, software, processed materials, or services. The supplier must establish the responsibility for each design including its development phase and verification using a formal design protocol. Design and development activities must be assigned to qualified personnel who are equipped with resources adequate to support the processes.

The design protocol requires documentation as the design evolves. The design plan should identify the organizational and technical interfaces among various groups, and information transfer among the groups is to be documented and reviewed on a regular basis using formal design reviews. The minutes of these review meetings become the quality records that document the design control process. The supplier must also identify and document all the design input requirements that will be reviewed and selected for adequacy. Incomplete, vague, or conflicting requirements shall be resolved by the team or group responsible for formulating the final design input requirements. This team or group is usually cross-functional and might be called a design review board.

The output of the design process must be documented in verifiable terms and include the design requirements, calculations, and analyses. The output shall:

- meet the design input requirements

- include or make reference to acceptance criteria

- conform to all regulatory requirements even if not stated in the input

- identify all design characteristics that are crucial to safe and proper function

Design verification must be planned, documented, and assigned to competent personnel. Design verification is to be used to determine that design output meets all of the design input requirements and is accomplished by methods such as:

- conducting and documenting design reviews

- prototyping

- computer analysis and simulation

- performing tests and demonstrations

- carrying out alternative calculations

- comparing a new design with a similar but proven design

The supplier shall establish and maintain procedures for the identification, documentation, and appropriate review and approval of all design changes and modifications.

DESIGN PHILOSOPHY

The design control section of ISO 9001 is structured to allow considerable flexibility in the design process but requires thorough documentation. An older, linear style design process can fit the ISO model and so can the newer techniques. Even though it is not the purpose of this book to explain or to teach design principles, the authors do wish to clearly illustrate the generic nature of the ISO guidelines by briefly examining modern design philosophy.

Concurrent engineering is a frequently used term referring to design by an interdisciplinary team that includes:

- customers or an adequate knowledge of their needs and desires

- design engineers

- purchasing expertise

- manufacturing/production engineers

- quality/test engineers

- service engineers (including field service)

- management/accounting expertise

The major goals of concurrent engineering are to shorten the design cycle, decrease costs, and increase value. If these goals seem to be at odds with one another, then perhaps you are only familiar with the traditional linear design process.

Figure 5-1 shows a linear design model. Although this model works and has produced quality products over the years, there are alternate ways of achieving design goals. Some of the potential problems with linear design include:

- a long design cycle

- designs tend to set up prematurely

- unrealistic designs

- expensive designs

- changes are costly and time consuming

- later design functions (i.e., manufacturing) are unnecessarily complex

- greater chance for errors of omission

- less than optimum use of available expertise

Figure 5-2 illustrates a model for concurrent design. Here, a team considers all of the design factors that affect quality, cost, performance, manufacturing, and servicing in a continuous improvement loop. ISO 9004 provides extensive checklists that can be used to guide a company in structuring the design review process:

a) customer needs and satisfaction

    1. comparison of needs expressed in the product brief with technical specifications for materials, products, and processes

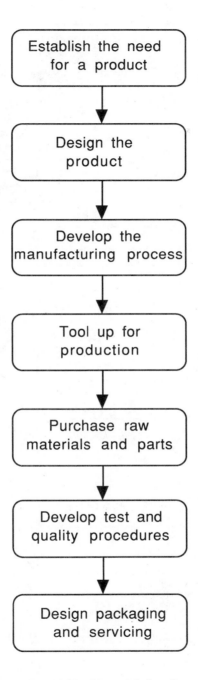

Figure 5.1 A linear design and development model

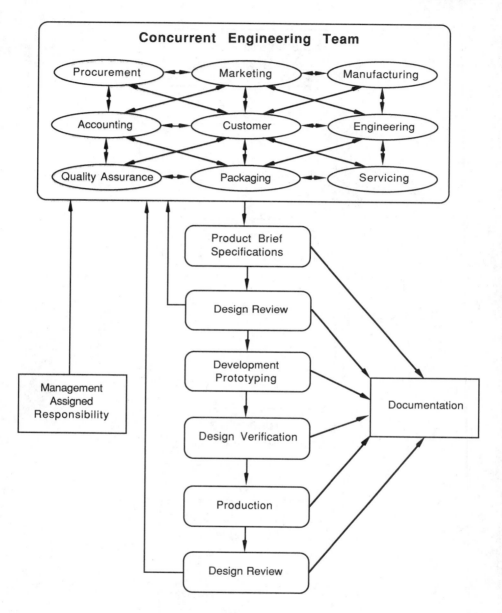

Figure 5.2 A concurrent design model

2. validation of the design through prototype tests

3. ability to perform under expected conditions of use and environment

4. considerations of unintended uses and misuses

5. safety and environmental considerations

6. compliance with regulatory requirements, national and international standards and corporate standards

7. comparisons with competitive designs

8. comparisons with similar designs and design history to avoid repeating problems

b) specifications and service requirements

1. reliability, serviceability and maintainability

2. comparison of tolerances with process capabilities

3. acceptance and rejection criteria

4. installability, ease of assembly, storage needs, shelf-life and disposability

5. benign failure and fail-safe characteristics

6. aesthetic specifications and acceptance criteria

7. failure modes, effects analysis and fault tree analysis

8. ability to diagnose and correct problems

9. labeling, warnings, identification, traceability and instructions

c) process specifications and service requirements

1. manufacturability, special processes and installation

2. inspection and testing requirements

3. material, component and sub-assembly specifications including approved suppliers

4. packaging, handling, storage, shelf-life and safety considerations

d) market readiness

1. availability and adequacy of installation, operation, maintenance and repair manuals

2. adequacy of distribution and customer service networks

3. training of field personnel

4. availability of spare parts

5. field trials

6. certification of qualification tests

7. inspection of early production units including packaging and labeling

8. evidence of process capability to meet specifications on production equipment

DOCUMENT AND DATA CONTROL

4.5 DOCUMENT CONTROL

Effective control of documentation is fundamental to the implementation of ISO 9000. It does not matter which standard is being implemented, 9001, 9002, or 9003; they all require strict adherence to good control of all documentation affecting either the product or service being offered.

Control of documentation is meant to insure that the correct information is available to support any process that could affect quality. Many have interpreted this to mean that there must be well documented manufacturing processes. This is true, but there must also be documentation of administrative processes as well. Up-to-date procedures are required in such non-manufacturing areas as:

• scheduling

• purchasing

• engineering

• human resources

• shipping

• warehousing

- packaging

- quality assurance

The standard also specifically states that any changes to procedures must be reflected in updated documents. This is meant to insure that as people informally improve their work habits, these creative changes are captured in the documented procedures. Keeping updated procedures insures that the written procedure and the actual practice are identical.

Includes Software

Computers are becoming more and more important in the fabric of American society. Small companies as well as large corporations now have complex systems of computers to help every facet of business. Computers have replaced the majority of paperwork in this new age of information (but, unfortunately, have generated some new paperwork of their own). Control of computer software is a part of the documentation control requirement.

Software that can affect quality must have the same precision and accuracy of information as all other controlled documents affecting quality. This has typically been one of the most difficult issues to resolve in operations using computers to control processes and take measurements. Types of software that require control are:

- testing programs developed in-house

- process software that requires set-point input

- product testing algorithms

This does not include purchased software that is not able to be modified by in-house personnel. Methods of control can vary. Software stored on main frames must have security inaccessible to non-authorized personnel. The software must also be titled, dated, verified for accuracy, and have revision numbers. A list of all changes must be documented and correlated to the changes made at each revision. If the software is stored on diskettes, the diskettes must be titled, revision labeled, dated, verified for accuracy, signed, and stored in a secure area. Backup media, such as tapes, must be handled in the same way.

The Requirements

The ISO 9000 standard for document control (Section 4.5) states: "The supplier shall establish and maintain documented procedures to control all documents and data that relate to the requirements of this International Standard . . ." It further states that these "documents and data shall be reviewed and approved for adequacy by authorized personnel prior to issue." Other requirements of document control are:

- The correct revision levels must be available to the people doing the work.

- Obsolete documents must be promptly removed.

- Documents must be changed to reflect process improvements.

- The nature of the change shall be identified in the document.

- A master list of all controlled documents must be maintained.

Keys to Implementation

One of the first decisions to be made about documentation is to centralize or decentralize the administrative control function. It may seem appropriate to let each area control their own documentation. However, experience has shown that centralizing the control mechanism uses the fewest overall resources. The basic goals are accurate documents, consistent documents and efficiency. The central documentation area handles the administrative functions of:

- insuring the correct format

- keeping track of revision levels

- maintaining the list of reasons for modifications

- managing the distribution list

- printing the document on the controlled paper or stamping it with the controlled stamp

- administering the temporary change procedure

- obtaining the correct signatures

- insuring the obsolete document is removed from use

Centralization does not mean that the document release process has to be cumbersome. The authorization process must come from appropriate personnel in each separate area. This can be done quickly if the individuals involved in authorizing documents are responsive to getting changes implemented quickly.

Procedure for Control

A key step in the documentation process is to write a procedure that defines how to write work instructions and procedures. Its main function is to set the format for all controlled documents. Items that should be covered are:

- document title
- document control number
- revision number
- name of approver
- date of effectiveness
- page number (e.g., 1 of 5)
- mode of control (red stamp, colored paper, and so on)
- revision control forms
- revision control history records
- informing those who need to know of revision
- control of public domain documents (e.g.: MIL SPECS)

Method of Control

A major issue to resolve is how controlled documents are to be identified. There are many ways to do this, but one of the most effective is to use red ink and a controlled stamp to identify all pages of all controlled documents to provide high visibility. The stamp might say "CONTROLLED DOCUMENT," "CONTROLLED DISTRIBUTION," or "CONTROLLED IF RED." This stamp gives evidence that the copy is an authorized, controlled copy. Therefore, if you find a copy with a "CONTROLLED DOCUMENT" stamp in black, you know that the document has been copied and may be out of date. If there is no stamp, red or black, then you know the document is not authorized and should be discarded.

Another effective method of control is limiting the type of paper on which controlled documents are printed. Color coded paper or specialty paper with a colored border can be used. Whatever the method, security of the stamps or the colored paper is a must.

If a temporary or emergency note must be distributed to the work force, then a form should be developed requiring a signature and date. The information on the form should result in a permanent change to a procedure, so the form must have an expiration date to insure that the information is captured in the permanent format (the procedure).

Security

Whatever method is used to identify controlled documents, there must be security to insure inadvertent use. If a stamp is used, it must be kept in a secure place such as a locked desk drawer or cabinet. If colored paper is used, it must not be available to people other than those in the document control group.

No "Post-Its" Allowed

The work force must be trained to recognize all documents without obvious evidence of control as *uncontrolled*. Any unsigned or undated memo taped to a cabinet, manufacturing machine, bulletin board, or office wall is an uncontrolled document and should be discarded.

Changes to Controlled Documents

If a quality system is to be effective, it must be easily changed to reflect improvements. A method of controlling changes must promote and encourage the ability to make improvements. If it is difficult to change documents, then efforts for continuous improvement will be more difficult. One way to make changes easy to implement is to develop a form called a *temporary change form* (see Appendix D). This form can be filled out by the person desiring the change and can be implemented immediately. The form must be authorized by the same level of person who authorizes procedures for the area. The form must have an expiration date (30 to 90 days is typical) to insure that the change becomes permanent if it proves to be a true improvement.

If a change must be implemented immediately, the form can be filled out, authorized, dated, and attached to the relevant procedure. An alert must also be sent out to inform everyone that the procedure has been changed (see Appendix D). Often it is required that all affected by the change sign the alert to insure that the change is uniformly implemented. These alerts are quality records and show evidence that all are informed of procedural changes.

A Disciplined Approach

Document control must be a disciplined process. There are many ways to insure that the up-to-date procedures, work instructions, prints, specifications and quality plans are at the point of use. Whatever method is used, it must be clearly defined and rigorously implemented if the ISO quality system is to be a success.

4.16 QUALITY RECORDS

This element of the standard requires the establishment and maintenance of procedures for records including identification, collection, indexing, filing, storage, maintenance and disposition. Records should demonstrate how the required quality is achieved and how the quality system operates. Relevant sub-contractor records are to be part of the data base.

The records must be legible and be identified with the product involved. They must be stored and maintained in a safe environment and in a readily available way. There should be an established procedure, described in the quality manual, regarding record retention time and disposition. They may need to be available to a purchaser's agent for an agreed time period.

ISO 9001 cites the need for quality records in the following areas:

Management Review	4.1.3
Quality Planning	4.2.3
Contract Review	4.3
Design Review	4.4.6
Sub-Contractor Assessment	4.6.2
Purchaser Supplied Product	4.7
Product Identification and Traceability	4.8
Process Control	4.9
Receiving Inspection and Testing	4.10.2.3
Inspection and Testing	4.10.5
Inspection Equipment Calibration	4.11
Test Hardware/Software	4.11
Nonconformity Review	4.13.2
Training	4.18
Internal Quality Audits	4.17

Quality records provide evidence that the quality system is working. The records will be scrutinized in a formal registration audit and will be thoroughly reviewed at each subsequent surveillance audit following registration. Experienced auditors realize that if a quality system is marginal or starting to deteriorate, it will show up first in the quality records.

There should be strict rules about how to properly fill out a quality record, how to revise an erroneous input, and how to maintain quality records for easy retrieval. Pencils must never be used because erasures might leave no evidence of changes. Corrected data must show evidence of the original data, the date of the correction, and the initials of the person who made the correction. This is usually accomplished by drawing a single line through the erroneous item so as not to render it unreadable. Then, the correct item is written above or below the lined item, initialed, and dated. Some companies prefer a specified color, such as red, for all corrections. Procedures such as these maintain the integrity and traceability of the quality records.

OPERATIONS CONTROL

4.8 PRODUCT IDENTIFICATION AND TRACEABILITY

Product identification is required for all components, subassemblies and finished product. The standard requires that "... the supplier shall establish and maintain procedures for identifying the product by suitable means from receipt and during all stages of production, delivery and installation...." Product identification can be accomplished in many different ways. The use of labels, stamps, tags, placards, and so on, all have proven to be effective. Your system is proven effective when all materials are identifiable at all areas of the plant. A simple universal tag attached by a string is one of the best ways to identify almost any type of product. The tag should denote:

- the product or part name

- the part number

- the lot number

- the container number

The advantage of using only one type of tag or label is that it is easy to spot when it is not there. When a company allows any type of identifying marking or tag, it can be very difficult to see when something is missing. If all personnel are conditioned to look for only one type of tag, then its absence will be obvious.

Traceability

If traceability is not contractually required by customers, it is not an absolute necessity of the standard. However, good traceability systems enhance the ability to properly address customer concerns, bracket nonconforming product, and conduct meaningful corrective action investigations in the support of continuous improvement. A world-class operation almost always has a system for reconstructing the production records back through all processes as well as tracking product forwards through all distribution channels to the customers. Although electronic lot records may be more efficient,fancy computer systems are unnecessary to have an effective lot traceability system. Some of the best traceability systems are based on non-computerized paperwork. The following description is for an actual system:

> All products can be tracked back to subassemblies and components with the use of lot files. Each product lot has a file containing all data generated during the manufacture of that lot. The data sheets are formatted so that all raw material lots or other component identifiers are written down on the production sheets. These production sheets are stored in the lot files and facilitate tracing all products back to the raw materials, operators performing the work, date of production, inspection results, process conditions and SPC data. Archive samples of the production are also retained to promote investigative work if a customer complaint should arrive. The only computer assistance is a very basic PC data base for keeping track of which file cabinet drawer the lot record is stored in. This enables the lot records file clerk to quickly obtain the paperwork necessary to investigate any traceability issues. This particular system requires significant floor space and two administrators to maintain it, but it performs admirably.

4.9 PROCESS CONTROL

The process control clause of the ISO 9000 standard requires the use of *controlled conditions*, including sufficient documentation, suitable monitoring, appropriate approval, and control of workmanship. Proper planning of production operations must insure that these proceed in the specified manner and sequence. Controlled conditions for process control include appropriate controls for materials, production equipment, processes and procedures, computer software, personnel, and associated supplies, equipment maintenance, utilities and environments. Processing operations must be specified to the necessary extent by documented work instructions.

Manufacturing Hub

Process control is the hub for:

- product identification
- process inspection and testing
- calibration
- test status
- nonconforming product control
- process quality records
- statistical techniques (SPC)

All of these elements of the ISO standard are related to process control in that they will usually be observed in the manufacturing area of plants. When auditors wish to review the above elements, they will go onto the shop floor to see them.

Typical Audit

Process control is the heart of any manufacturing operation. It is so all-encompassing that a registration audit will focus much of its resources on investigating process control. A typical audit could go something like this:

An auditor walks up to an operator and asks what she is doing. The operator refers to her production schedule and reports that she is working from these instructions. The auditor looks at the schedule and asks the operator how she knows that this is an up-to-date schedule, how it is authorized (signatures and dates) and if it ever changes. If it does change, who is authorized to change it and where does the authorizing signature and date appear? Once that has been established, the auditor might ask the operator for a written procedure that authenticates the just-discussed verbal procedure. The operator must be able to produce the procedure in a reasonable amount of time (a couple of minutes) to show that the written procedure is readily available (4.5.2). The auditor will inspect the document to insure that it is a controlled copy and a current revision. It will also be checked for unauthorized revisions in pen or pencil. The procedure will be scrutinized to see if it matches with the operator's verbal discussion. Any place a setpoint, specification, or process condition is referred to, it must show a process range rather than a single target value. For example, if the procedure states that the temperature of water cooling a vessel must be 65 degrees, the auditor could ask "What if the temperature is 64 degrees?" The operator might say "Oh, that would be OK; it is close to 65 degrees." The auditor could then ask "What if it is 50 degrees?" The operator would probably say "No, that would be too cold." Then, the auditor asks "How about 57 degrees?" As you can see, set points and specifications must be stated in such a way that all operators will interpret the procedures in a uniform fashion. This is the only way operator variation can be controlled and quality maintained. The best way to maintain consistency is to define a process range, perhaps 60 to 70 degrees, if that is reasonable and controllable. The auditor could then go to some boxes being staged in the area and ask how the operator knows what they are. The operator might point to a label or tag and say that is how she identifies the product. The auditor could ask if the operator ever gets any bad raw material or if she ever processes material that is not considered good. Often, the operator will point to some product that is being held because it has not passed the required quality checks. The auditor could then ask how the suspect product is segregated from the good product. What the auditor expects to hear is that the product is labeled as nonconforming and stored separately. If the auditor finds nonconforming product without a nonconforming label or not segregated from good product, the auditor has cause to write a noncompliance. The auditor can also ask if nonconformances are tabulated and reported so that corrective action can reduce or eliminate recurrence.

Process control usually relies on measurements taken at reasonable intervals. In-process inspections are usually the vehicle for these process measurements. An auditor can ask an operator if he takes measurements of a process. If he says yes, the auditor can ask to see him take some. An auditor can also ask to see the documented quality plan that defines the appropriate techniques for taking these measurements. These documents would be checked for the proper revision level, appropriate authorizing signatures and for visible signs that they are properly controlled documents. The auditor can check the measurement equipment to determine if it is properly labeled with calibration information. The label should have an identifying number, the next calibration date (hopefully beyond the current date), and an authorizing signature.

Production Equipment Too

Production equipment must also be controlled and maintained to insure adequate process control. There must be maintenance records to show that the equipment is properly maintained. When special processes are involved (see below), equipment must be certified to insure that it can consistently make good product. The certification process must be thoroughly documented to show that the operating conditions approved for the equipment result in the production of good product. Environmental conditions for the equipment must also be taken into consideration.

Records

Auditors look for evidence that the quality systems in place are actually being used on an ongoing basis. In order to investigate this requirement, they ask to see testing records, calibration records, training records, nonconforming product reports, corrective action reports, and minutes of meetings discussing continuous improvement efforts for the area. Unadulterated records are essential to demonstrate the proper use of the documented quality system.

Special Processes

ISO 9001 notes special processes as those processes which can not be fully verified by subsequent inspection and testing of the product and where processing deficiencies may only be detected after the product is in use. ISO 9001 requires that special processes be continuously monitored with rigid compliance to documented procedures to insure that the specified requirements are met. Special processes might include welding, soldering, painting, heat treating, and so on.

Special processes must show evidence that the approved processing conditions result in acceptable product. Usually a process certification study is conducted so that measurable process conditions can be correlated to product characteristics that cannot be easily inspected. Once the study has verified the appropriate processing conditions, the process parameters must be monitored and recorded to show evidence that the resulting product is acceptable. If process upsets occur, special steps must be taken to insure that the product is still acceptable. This might include destructive testing of a statistically valid sample of the suspect product. It is very difficult to have a valid control scheme for special processes without using statistical process control (see Element 4.20).

Process control is one of the key elements of ISO 9000 and one which will be closely scrutinized in a registration audit.

4.10 INSPECTION AND TESTING

Inspection and testing incorporates all of the verification activities for materials from incoming raw materials to final product. Verification is the operative word since the product need not be inspected for it to be verified as acceptable. The method by which product is verified must be documented with quality plans or other documented procedures.

Receiving Inspection

Controlling incoming materials is usually coordinated closely with the purchasing department. The purchase order or contract with the supplier should specify how raw material is to be verified as acceptable. Many progressive companies have pursued partnerships with their suppliers and have developed good working relationships that minimize or eliminate the need for costly incoming inspection. This can be accomplished by using supplier audits or some other form of supplier certification. For those suppliers passing the requirements, and demonstrating their ability to consistently deliver good products, there should be no need to inspect their shipments to verify quality. Good record-keeping is essential to verify incoming product in this manner. There should be an evaluation system that rates suppliers and monitors their deliveries over time. This is important because in these turbulent times, good suppliers can turn bad with management turnover or hostile takeovers. Suppliers must know that their offering is being monitored and that if they do not perform, they will be replaced. It is also in everyone's best interest if regular reports are published showing performance ratings of suppliers. It not only summarizes the verification activities for incoming product, but it also encourages suppliers to continuously improve their offering.

Inspection is Necessary

Unfortunately, many firms still meet this ISO requirement by inspecting a large quantity of incoming product. If incoming inspection is used, procedures or quality plans must be documented defining in detail how the material is to be inspected. These quality plans must define:

- the measurements to be taken

- the equipment to be used

- the specifications

- the frequency of testing

- record-keeping requirements

- labeling requirements

- nonconforming material procedures

Comprehensive records of the measurements taken must be maintained as evidence that incoming product was verified. The records should show:

- the supplier

- the product inspected

- the measurements taken

- the measurement values observed

- if the values are within the required tolerance range

- any required calculations

These records will be used to evaluate a supplier's performance and will be the primary source of information in the supplier rating system.

In-Process Inspection

In-process inspection is basically the same as measuring the process. It can be accomplished by measuring the product that results from the process or measuring process conditions that help control the product being manufactured. The results of the measurements are used to not only keep the process under control, but to look for opportunities to make improvements.

Process reports should be compiled and analyzed by process techni-cians or engineers to see if trends are evident. If these analyses are used in the right way, it will keep improving the process and should result in minimizing customer complaints. In-process inspection is basically the measurement portion of process control (see Element 4.9). Once the mea-surements are taken, statistical process control should be used to translate the raw data into meaningful information (see Element 4.20).

Final Inspection

Final inspection includes not only making final measurements of the product before it is sent to the customer, but also includes verification that all previous tests were performed and that all specifications were satisfied. In some firms this is accomplished by clerical compilation of all pertinent data into batch or lot records. These records are extremely important if traceability is a requirement.

Inspection and Test Records

When people think of quality records, inspection records are the first thing that comes to mind. Any place where a measurement or check has taken place should be recorded and included in the inspection and test records. These records must not only show evidence that the product was good, but it must also show when the product or process did not meet specifications and what actions were taken to rectify the situation. As with all quality records, signatures authorizing further processing or shipment to customers must be in evidence. There must be a trail back to the autho-rizing document showing the authority for release of product to the next process or to the external customer.

4.11 INSPECTION, MEASURING AND TEST EQUIPMENT

Control of inspection, measuring and test equipment is a crucial and sometimes difficult item in an ISO 9000 implementation.

Comprehensive List

All measurement equipment that can be used in a way that will impact quality must be identified to insure a comprehensive list. Personal gauges such as micrometers used in any measurements where quality can be impacted must be included on the list. Even jigs, fixtures and templates used for making measurements and quality decisions must be on the list. The list should include the location of each device, the frequency of the calibration, the method of calibration, and the acceptance criteria. A calibration procedure for each type of device is obviously warranted. Instruments and gauges used only for monitoring purposes do not have to be in the calibration program.

Traceable Standards

In order for the calibrations to be valid, they must be done with test standards that are traceable back to the National Institute of Standards and Technology (NIST). If calibrations are contracted out, there must be a calibration certificate with the appropriate calibration information included (test procedure number, NIST traceable standard number, and so on). The supplier of calibration services must be on the approved suppliers list (see Purchasing, Section 4.6.2). Where there is no NIST or other nationally recognized standard available, the basis for the calibration must be established. This would most appropriately be done with a statistical study showing the accuracy and precision of the data certifying the measurement device for use.

Identification

Once the list is complete, each piece of measurement equipment must be identified with a label showing (see Appendix E):

- the equipment number

- the next required calibration date

- an authorizing signature or a stamp imprint

Stamps, if used, must be carefully controlled. Equipment that might be used to make quality decisions, but is not, can be identified with a "NO CALIBRATION REQUIRED" tag. Examples include such items as flow gauges, pressure gauges, and thermometers used to identify the presence or absence of materials or conditions. This might seem like overkill in labeling, but it does help to differentiate between gauges requiring calibration and those that do not.

Measurement Capability

When a measurement device is used to take a measurement, it must have the correct accuracy and precision to do the job. The thickness of a human hair cannot be measured with a yardstick; the speed of a sprinter cannot be gauged by a sundial; or the distance to the moon cannot be determined with a ruler. The intended measuring instrument must be capable of resolving an increment that is at least 4 times smaller than the specification of the item being measured (some say it should be ten times smaller). For example, if wooden dowel rods measured 48 inches, plus or minus 1/2 inch, then the specification band is 1 inch in length (47.5 to 48.5). To meet the four-times-smaller requirement for the measurement device, the smallest increment for the tape measure would need to be 1/4 inch. To meet the ten-times-smaller requirement, the tape measure would need to be incremented to 1/10 of an inch. Where warranted, measurement capability studies should be conducted to insure the device is appropriate for the measurement.

Recall System

There must be a system by which measurement devices are recalled for calibration. One possibility is to use a personal computer spreadsheet with listings for:

- the device number

- the device type (micrometer, caliper, and so on)

- the device location

- the device owner

- the last calibration date

- the calibration due date

A calibration technician can access the spreadsheet and request a report sorted by calibrations due that week. The technician can then send out a form letter requesting that the devices due for calibration are made available to the calibration department. Any devices not found or unavailable can then be listed on another report (addressed to management) of delinquent calibrations.

Secure Environment

The calibration laboratory must be secure from unauthorized personnel and maintain the proper environment to insure accurate and precise calibrations. It is generally unacceptable to perform calibrations in manufacturing areas with extremes in temperature, humidity and cleanliness (or lack thereof). Obviously, these are not good conditions for achieving highly precise and accurate calibrations. Also, there should be restricted access to the calibration equipment. Tampering with precise equipment by unauthorized personnel can compromise the calibration system.

Test Result Validity

When a measuring device is out of calibration, there must be a procedure for addressing the situation which could impact the quality of the product shipped to customers. If erroneous results are accepted as valid, out-of-specification materials might have been shipped. The impact of this situation must be studied and a potential recall of the material must be considered. The seriousness of the situation can be assessed only if the calibration values achieved are documented. This is another reason why quality records are so important. Without the raw data for analysis, how can a proper judgment be made on the best course of action?

Software Too

If test software is used in the measurement process, it, too, must be checked to insure acceptability of the product and then rechecked at some regular interval. If the master copy of the software is stored on diskettes, a similar label as the one used for test hardware can be used. Security must also be maintained on the test software.

4.12 INSPECTION AND TEST STATUS

Inspection and test status are often confused with Element 4.8, identification and traceability. Confusion occurs easily because both requirements often involve a tag or label for compliance. Therefore, complying properly with product identification and traceability should enhance the compliance with inspection and test status.

The primary concern with inspection and test status is that nonconforming material is not confused with good material. If control of nonconforming material (Element 4.13) is done properly, this should not be a problem. However, there may be other possibilities needed to differentiate quality status. Product can sit in a production area with the following status:

- awaiting incoming inspection
- experimental product
- hold for engineering analysis
- quality control sample
- awaiting final inspection
- hold for 100% inspection

Product in all of these situations must be differentiated with proper identification. This can be done with a special "HOLD" tag requiring a written description explaining the reason the material is being held.

In manufacturing operations, where a series of process steps are required and the product appearance does not significantly change from process step to process step, it is important to keep track of the status of the product at all times. An excellent tool for this purpose is *a traveler* (lot traveler, material traveler, process traveler, and so on). As each process is satisfactorily completed, an operator signs and dates the traveler to provide authorization to allow the product to pass on to the next step. Once the lot travelers are completed, they can be retained as quality records to prove the satisfactory completion of process steps and quality checks (see Appendix E).

An electronic traveler, created with a computer program using barcoding, can be a very useful tool as well. Whenever an operator completes an operation, he must wand the product tag bar code. Anyone with access to the computer system can then determine the status of all product in the facility. These systems are normally a key part of an effective MRP system (Materials Requirement Planning, see Chapter 6). Electronic records provide evidence that the appropriate manufacturing steps have taken place.

4.13 CONTROL OF NONCONFORMING PRODUCT

Control of nonconforming product is meant to insure that any product unable to meet specifications is precluded from use. The control requires a system provide for:

- identification

- documentation

- evaluation

- segregation

- disposition

- notification

Documented procedures are required to define the process by which the data pertaining to the nonconforming material is to be reviewed. Many customers require notification of product that does not conform to specifications. Nonconforming product must not be reworked and shipped without informing them, and all data pertaining to the nonconforming product must be made available to them.

Identification

A nonconforming product must be identified in some way. There are many viable techniques available, but one of the most effective is the use a large, red label or tag with the words "NONCONFORMING PRODUCT" written on it. Another method includes placing the product on a shelf, in a container, on a table, on the floor, in a cage, or in some other area labeled "NONCONFORMING." This option is less effective because good product can be readily mixed with bad product.

Documentation

To protect good product from bad, efficient documentation is essential. This form of protection, a *nonconforming product sheet,* not only documents a nonconformance, but it also initiates the corrective action process (see Element 4.14). Without documentation, the bad product cannot be properly segregated from the good. Documenting nonconforming product also develops a data repository for tracking continuous improvement activities.

In the past, nonconformance reports were feared and dreaded. Old-style management often reacted negatively to nonconformance reports by trying to place blame. Emphasis was not on continuous improvement, but instead on who errored and how punishment would be administered. Today, nonconformance reports are considered opportunities for improvement.

Evaluation

All data pertaining to nonconforming material should be evaluated for any evidence of recurring problems. These preventive measures are required in the ISO standard and present the opportunities for improvement.

Segregation

It is a good operational practice to set aside special areas for nonconforming product. Many manufacturing areas are disorganized, and storing good product near nonconforming product runs a high risk of inadvertent mixing. This is a recipe for disaster. Without control, nonconforming product appearing on the documents and the scrap report might be found in a customer's order. It is very difficult to convince customers that a company has a state-of-the-art quality system when scrap product shows up on their receiving dock. It is also tough to convince a registration body that the quality system is worthy of ISO 9000 registration.

Disposition

Product identified as nonconforming must be dealt with in some manner. There should be very explicit rules for processing procedures. Potential options include:

- scrap it

- rework it

- reinspect it (100% inspection)

- use as is

Deciding what must be done with nonconforming product is not a job for any employee or group who has much to gain and little to lose by using product as is. Normally, disposition decisions are handled by a Material Review Board (MRB). The board is made up of marketing, engineering, quality, purchasing, design, and technical representatives who can best assess the impact of the nonconformance on the customer. A representative from the MRB should be responsible for alerting the customer to the "ship as is" decision and to insure corrective action is initiated to prevent recurrence of the nonconformance. Some companies use an Engineering Control Board to perform this function.

If product must be scrapped, there should be a process available by which the scrap product is sent to a dumpster or another proper repository so it does not find its way back into production. One successful method of control is to collect scrap disposition tickets. These tickets are signed off by the person who physically scraps the product and they are then retained as a part of the production records.

Rework is a process by which the product is made whole by some extra post-manufacturing step. This is an option that can create problems if not properly controlled. Any reworked product must pass all requirements imposed on good product. Reworked product requires a complete inspection that is at least as rigorous as that used for normal production.

Reinspection is normally initiated when a high percentage of visual defects are found. A 100% inspection avoids scrapping the entire lot of production, but it can become expensive and far from fool-proof. It is widely accepted that 100% inspection is only 80% effective, so there are no guarantees that all product will be good. Customers should be alerted to the requirement for reinspection so they can be on the lookout for defects that passed through. This is not a popular step with marketing, however, because anything interfering with customer confidence in product quality is unappreciated. For that and other reasons, rework, reinspection and use-as-is processing must be tightly controlled.

Notification

Any time nonconforming materials are made or detected, all parties affected by the nonconformance must be notified in writing. This applies to internal as well as to external customers. When processing and documentation costs are added up, it is easy to see that nonconformances are extremely expensive. When a nonconformance occurs, it should not be just sloughed off as a part of doing business, but should instead be taken seriously. If the business is to remain successful, aggressive action must be taken to reduce or prevent recurrence.

4.15 HANDLING, STORAGE, PACKAGING, PRESERVATION AND DELIVERY

Section 4.15 is intended to insure that all materials (including raw materials, in-process materials, and especially final product) are identified, protected from deterioration, segregated from defective material, and properly handled to prevent damage.

Final Product

In terms of final product, it is important to insure that, after significant time has been spent controlling all aspects of a product or service, no mistakes are made in the final stages prior to customer acceptance. The maximum value of the product or service is attained at the time of delivery, but often few resources are deployed to insure that the product is protected all the way to the customer. This must not be an afterthought but a carefully planned and controlled process.

Expiration Dates

Documented procedures are required for all administrative processes. Deterioration of product or materials with expiration dates must be prevented. Procedures barring expired product or material shipment must be observed.

Storage and Shipping

Storage and shipping can be sources of trouble unless they are tightly controlled. Product and materials can be damaged by fork lifts, smashed by improper stacking, contaminated with water, dirt or chemicals, or subjected to extreme temperatures or humidity. If warehousing and shipment are contracted, it is very important for the contractor to know and meet the requirements of the product. Word-of-mouth instructions are not enough. Documented procedures are essential for legal and regulatory purposes, and they are a good business practice. Outside providers of warehousing and shipment services must be on the approved suppliers list (see Purchasing, 4.6.2).

Packaging

Packaging is often an afterthought when it comes to funding quality initiatives. The first thing a customer sees when using a product is the packaging, and poor packaging materials and practices can damage a company's image. The old adage about first impressions certainly applies here. The product may be great, but if the packaging fails to protect it or if the packaging is "user hostile," the customer will view it as less than desirable.

Continuous Improvement

4.14 CORRECTIVE AND PREVENTIVE ACTION

Corrective and preventive action is a vital requirement for the prevention of nonconformances as defined in Section 1 (scope) of the ISO 9001 standard. Corrective and preventive action is the cornerstone of the ISO standards. It is the one sure way (if properly managed) to pay for the resources required to implement a quality system that fosters continuous improvement. Without effective root cause analysis and implementing effective corrective and preventive actions, old inefficient operations may remain. Corrective action can be initiated from a number of potential sources:

1. customer complaints

2. supplier problems

3. internal audit results

4. internal processing problems

Each of these problem sources or nonconformances must be addressed in corrective and preventive actions if the quality system is to function properly.

The corrective and preventive action section requires an organized system to:

- investigate causes so as to prevent recurrences

- analyze all processes and work operations to detect and eliminate potential causes of defects

- initiate preventive actions

- insure that corrective and preventive actions are effective

- implement and record changes resulting from corrective and preventive actions in procedures and work instructions

Customer Complaints

Most operations treat customer complaints with a specific procedure, and most companies recognize this as a key to their survival. Customers will not tolerate repeat defects and they want action taken when substandard products or services are received. Unfortunately, many customer complaints are not evaluated for root cause, and consequently, problems are not solved. Too often, the root cause is identified as operator error, and the corrective action is prescribed as "instructed operator in proper procedure." Often, this action is not based on a proper analysis and, therefore, is not an appropriate corrective action for many problems. Frequently, operators make mistakes because a system is poorly conceived and prone to error. Most operators want to do a good job, but sufficient resources (tools, time, paperwork, equipment, and so on) might be lacking to do the job correctly and effectively. It is management's responsibility to provide the proper resources. Corrective action can be the vehicle by which management is informed of the need.

Processing Problems

Nonconforming product or nonconforming process conditions can be sources for corrective actions, but not all occurrences need to be investigated. Judgment needs to be exercised when deciding to undertake a corrective action. The cost of the nonconformance, the number of repeats of the same nonconformance (whether or not it is due to assignable causes), and the potential customer impact if the nonconformance had not been detected are some measures by which corrective action might be required. In all cases, a conscious decision must be made whether to investigate with a full-blown corrective action or whether to correct just the symptom and log the occurrence for further analysis later.

Supplier Problems

Supplier problems are often addressed through the purchasing function. A good corrective action system for suppliers fosters a team approach between purchasing, quality assurance, manufacturing, and engineering in developing partnerships with suppliers. The purchasing department also benefits because purchasing personnel will have a better link to the needs of manufacturing and, therefore, develop better working relationships. Purchasing is looked upon as a valued asset to the manufacturing organization rather than a "buddy" of the supplier. A corrective action system for suppliers brings the purchasing department into the mainstream of the quality system.

Systemic Approach

Corrective and preventive action investigations must strive for a systemic correction. Addressing the symptom will not correct the disease nor prevent recurrences. Rarely are repetitive problems truly caused by operator error. For example, if operators are making excessive nonconforming material, perhaps the root cause is due to problems such as:

- substandard raw material
- excessive machine wear
- improper operator training
- poor material feed systems
- too many non-value added steps (excessive handling)
- no preventive maintenance system
- poor equipment design
- understaffing of the operation

Once the real root cause has been identified, the corrective and preventive action must address the problem at its source, such as:

- poorly defined material specifications
- understaffed maintenance
- inefficient training programs
- poorly designed material feed systems
- inefficient process flow
- lack of manufacturing involvement in equipment design
- short-term focus on cost containment (lean and mean)

Substantial courage is often required to inform management of their cost containment efforts which are resulting in excessive operational costs. However, there is a better chance of correcting the true cause if the discussion is backed up with data and statistics.

Follow-up System

In order for the corrective and preventive action system to be effective, there must be a monitoring process by which open corrective actions are expeditiously completed. This system needs triggers for key steps in a corrective and preventive action process:

- corrective action meeting held
- minutes of corrective action meeting distributed
- corrective and preventive action analysis completed
- corrective and preventive action plan completed
- corrective and preventive action steps implemented
- corrective action verified for effectiveness

This can be done with a manual system or a computer spreadsheet. Notices should be distributed when a due date for one of the above steps is passed without action being taken. Copies shared with management help insure that corrective and preventive actions do not languish or prove ineffective.

4.17 INTERNAL QUALITY AUDITS

Internal quality audits keep the quality system fresh and alive. For manangement, they insure the quality policy and quality objectives are being met. Internal quality audit results are a main topic in management reviews (see ISO 9001, Section 4.1.3). Records of quality audits will be closely scrutinized by registration body auditors when they return for surveillance audits.

Formal System

Documented procedures are required to define how the quality audit system is to operate. A schedule of audits insures that all important areas are covered with an appropriate frequency. The standard does not define what that frequency should be, but the accepted practice is for all areas to be audited at least once per year. Areas more critical to the operation of the quality system might need to be audited once every six months or once per quarter. Of course, these frequencies might need to be adjusted based on the historical results. Where many problems are encountered, the frequency of audits might have to be increased, while in trouble-free areas, the frequency can be reduced. One could probably make a case for one or less audits per year if the results justify it. However, in the beginning, annual audits for every area is recommended. An example of an audit schedule can be seen in Appendix D.

Training

Training auditors is a key issue for internal audits. Since this is a very critical area in the quality system, auditors must be qualified "... on the basis of appropriate education, training and/or experience, as required" (see ISO 9001 Section 4.18). Not everyone can become an effective auditor. Auditors must have inquisitive and analytical minds. They must exhibit strong interpersonal skills and be able to give bad news to a hostile audience without creating a riot. Auditors must be able to extract information by both listening to interviewees and by reading complex documents while on a hot, noisy, and dirty manufacturing floor. Auditors must have a great deal of self-confidence and not be intimidated by forceful auditees. However, auditors must not be stubborn and unwilling to listen to other often valid viewpoints.

Auditing is perhaps best done by a few, select, highly skilled people. Some firms choose to involve many people in the auditing process so that no one carries the burden constantly. These firms believe that employees can learn from what they see in an area foreign to them. However, the audit process is too important to give to amateurs. Ownership of the audit process by a few skilled auditors usually yields more professional audits and better promotes the continuous improvement process.

Independence

Auditors are required to be independent of the areas being audited. If the auditors are from the quality department, they cannot audit the quality department. Although difficult for some firms, this situation can be handled by using another auditor from a sister site or by hiring an independent auditor from the outside. This is an absolute requirement of the ISO standard (ISO 9001, Section 4.17, Paragraph 2).

Costs

Training costs can be prohibitive if too many people are involved. A good auditor training course can last from two to five days based on the degree of proficiency desired. Outside courses can cost in the range of $900 for a two day course to $2400 for a five day course (not including expenses). By choosing to train a large number of auditors at a company facility, money can be saved.

Records

Records for the internal auditing process should consist of:

- audit schedules

- memos to auditees announcing time and date

- opening meeting notes, where appropriate

- written noncompliances

- corrective actions taken for each noncompliance (including verification of implementation and effectiveness)

- audit reports to responsible management summarizing the results of each audit

- internal auditor training records

Audit results must also show up as topics of discussion in management reviews. Registration auditors will look for improvements to the quality system by reviewing the minutes of management reviews and the effectiveness of the internal quality audits as evidenced by the records.

Audit Records

Audit results must be acted upon by those responsible for the area being audited. Area management must have a system in place to address the deficiencies. Corrective actions must be taken to resolve the audit findings, and the corrective actions must be effective in preventing recurrence (see Element 4.14).

4.20 STATISTICAL TECHNIQUES

There is no absolute requirement to use statistical techniques in order to become ISO 9000 registered. However, auditors look for the use of statistical techniques to promote process control, conduct root cause analyses for corrective and preventive action, assess suppliers, conduct calibration control studies, and to convert raw data into more useful information.

Seven Basic SPC Tools

The most commonly used statistical tools are:

1. Pareto Diagrams

2. Flow Charts

3. Run Charts

4. Fishbone Diagrams

5. Control Charts

6. Correlation Diagrams

7. Histograms

These seven basic tools are simple to use and can be very powerful in transforming data into useful information.

Use It Correctly

Statistical Process Control (SPC) can be very effective but only when properly applied. Unfortunately, there are common abuses that make SPC useless and even misleading. The most frequently abused tool is the *control chart*. An auditor may ask why SPC is being used in the first place. If the auditor is told that the customers are requiring it, then skepticism grows. Using SPC to merely satisfy customers will probably not make a good impression on the auditor. The proper use of SPC promotes continuous improvement. This will manifest itself first in the choice of key parameters controlled with a control chart. Every process has a number of possible parameter choices. It is prohibitive to make control charts for all parameters, so choices must be made. The control chart should monitor key parameters and give a signal when the process is out of control. It is important to choose the key parameters in an appropriate way. Since the control chart is supposed to control the process, parameters that most frequently upset the process should be the ones chosen for control. Often, however, parameters that are the easiest to control are chosen because the objective of the control chart to appease customers is incorrect. In such a case, true process control is not taking place.

The Proper Way

A valid way to choose the key parameters is to use a Pareto Chart of defects or defectives to see which parameters cause the most scrap, rework, or customer complaints. The top two or three parameters on the Pareto Chart should be chosen, and more than three or four parameters become prohibitive to control.

Once the key parameters are appropriately chosen, control limits must be calculated. The control limits must be calculated when the process is in statistical control or when there is no assignable cause variation. In many cases, the control limits are inappropriate for the process. This could be due to one of the following:

- The control limits were arbitrarily set as some percentage of the specification (pre-control).

- The process was out of control when the control limits were calculated.

- The process has changed (improved or degraded) since the control limits were calculated.

Whichever is the case, a control chart with inappropriate control limits is not useful. It usually leads to erroneous process decisions. This can result in doing nothing when action is needed, or taking action when none is required. Either situation results in increased variation rather than a reduction, as desired.

One reason why control limits are assigned as a percentage of the specifications is for a preliminary gathering of data. This is not a problem unless the control limits are used to try to control the process. An unfortunate situation arises when the control limits are never truly calculated from a stable process. If false control limits are allowed to stay on the control chart, they do more damage than good because they give the false illusion of control. Then, process improvements do not happen, and people can lose their confidence in SPC and abandon it.

Another common occurrence is when legitimate signals occur, nothing is done. This often happens when the measured parameters are still inside the specification. Most operators and technicians are hesitant to shut down a process when the product being made is within specified limits. If the process shows a definite parameter drift but is allowed to continue running, knowledge is lost by not investigating the assignable cause of variation. This lost knowledge can translate into lost revenue. Even worse, money is wasted when processes are allowed to run out of control and nothing is accomplished by the use of the control chart. If they are not going to be used, the cost of generating control charts is akin to throwing money away.

Recalculate Control Limits

Control limits must be evaluated regularly. If control charts are used correctly, assignable causes of variation are investigated and removed. This dovetails with the corrective action portion of the ISO 9001 standard which includes investigating the cause of nonconformities relating to product, process, and the quality system, recording the results of the investigation, and applying controls to insure that effective corrective action is taken.

Once assignable causes of variation are permanently removed, the result should be reduced variation of the process. Defining when a process is in statistical control is not an exact science. Different rules are used. For example, many companies check a process when seven or more consecutive measurements fall on the same side of the center line (even when all of them are well within the control limits). The control charts will show that tighter control limits are required when the data resides in a very tight band around the goal. When this happens, a lack of recalculation of the control limits will result in losing the opportunity to make additional improvements. On the other hand, striving to make a process much better than it needs to be can become an interesting but impractical challenge.

Chapter Review Questions

4.1 Management Responsibility

1. What is the most important factor for success of the ISO implementation effort?
2. What is the top (key) document in an ISO 9000 quality system?
3. What must be included in a quality policy?
4. How might the quality policy be communicated to the organization?
5. Who in the organization must have their quality responsibilities defined and documented?
6. List some resources that must be supplied by management.
7. True or False: The highest ranking member of the quality department must be appointed as the quality system management representative.
8. How often should management reviews be held?
9. Who should attend management reviews?
10. List some typical agenda items for management reviews.

4.3 Contract Review

11. Why is it so important for the contract review process to be tightly controlled and thoroughly documented?

12. List some functional groups involved in contract review.
13. What form can contract review records take?

4.5 Document Control

14. To be effective, where must documentation be kept?
15. True or False: Manufacturing process documentation is the most important documentation to keep under control.
16. Name three types of software requiring document control.
17. List the types of information required on software diskettes.
18. List four of the five main elements of document control.
19. List some duties required of document control personnel.
20. List six items that should appear on all controlled documents.
21. Who must approve changes to controlled documents?

4.6 Purchasing

22. What are some key pieces of information that belong on all purchase orders?
23. List four ways that an organization can be impacted by controlling its suppliers.
24. How does variation in incoming materials affect automated equipment?
25. List some commodities or services whose suppliers should be controlled.
26. List some commodities whose suppliers do not need to be controlled.
27. What are some ways to measure suppliers' output?

4.8 Product Identification and Traceability

28. Name some ways products might be identified as they are processed through a factory.
29. When is traceability required?
30. What are some items that should be placed on identification tags?
31. How does traceability promote quality?

4.9 Process Control

32. Describe some controlled conditions as defined by process control.
33. List some of the ISO elements directly related to process control.
34. How are special processes different from other processes that must be controlled?

35. What are the requirements for the approval and maintenance of equipment?
36. Name some typical special processes.
37. What forms do process control measurements take?
38. How does statistical process control help to promote the control of special processes?

4.10 Inspection and Testing

39. True or False: In order to meet the requirements of ISO 9000, incoming inspection is required of all supplies that can impact quality.
40. List six items that good quality plans contain.
41. In-process inspection should use what quality tool to translate raw data into meaningful information?
42. What must be done at final inspection to insure that all testing has been completed?
43. True or False: Quality records must show only that product shipped to the customer met the required specifications.

4.11 Inspection, Measuring and Test Equipment

44. List some types of measurement equipment that are subject to the requirements of inspection, measuring, and test equipment.
45. What are some identifiers that must be associated with each piece of measurement equipment?
46. True or False: Suppliers of calibration services belong on the approved suppliers list.
47. True or False: If no national standard exists, it is permissible to do a study to define the standard internally.
48. True or False: All gauges used in a plant must be a part of the calibration system.
49. For measurement capability, the specification should be at least _____ times greater than the smallest increment on the scale of the measurement device.
50. Name five identifiers required on a comprehensive listing of all measurement devices in an effective recall system.
51. True or False: Product made while using an out-of-calibration measurement device or system must be recalled.

4.12 Inspection and Test Status

52. What is the primary concern of inspection and test status?
53. Name four different production status types.
54. What is an effective way to keep track of product as it is passed from one process step to the next?

4.13 Control of Nonconforming Product

55. Name the six system elements that must be defined in association with nonconforming product.
56. True or False: The only way to identify nonconforming product is to apply a "nonconforming" tag to each nonconforming item.
57. What is the most important use of nonconforming product documentation?
58. Why is segregation of nonconforming product so important?
59. What are the four possible disposition categories?
60. Who must be notified of disposition decisions?

4.14 Corrective Action

61. Name the four primary sources for corrective action activities.
62. Name the five elements of a corrective action investigation.
63. True or False: The majority of the time, the root cause of problems is operator error.
64. What are three factors that should impact the decision to investigate a problem with a corrective action investigation?
65. Which functional groups should be involved in a supplier corrective action investigation?
66. Name four possible reasons why operators could be making nonconforming materials.
67. Name the six key steps in a corrective action process.

4.15 Handling, Storage, Packaging, Preservation and Delivery

68. Why is it so important that product be carefully handled before being shipped to the customer?
69. What makes packaging such a key issue with customers?
70. Why must product storage be so well controlled?

4.16 Quality Records

71. List six examples of quality records.
72. True or False: Quality records must be kept in a central location.
73. What are the key elements of the control of quality records?
74. True or False: Paper copies of records are the only proper form for quality records.
75. True or False: Quality records must have the same control mechanism as all other quality documentation.

4.17 Internal Quality Audits

76. How does upper management use internal quality audits to insure that the quality system is being maintained?
77. What is the minimum recommended frequency for internal quality audits for each area?
78. Name four important characteristics for an auditor.
79. List the six major elements for internal audit records.

4.18 Training

80. What are the two main issues in the requirements for training?
81. Who must receive training in order for the ISO training requirements to be met?
82. True or False: The personnel department must retain all records of training for all personnel.
83. True or False: On the job training does not require formal training records.

4.19 Servicing

84. True or False: Servicing is a requirement for all those who make a shippable product.
85. True or False: Since servicing takes place at customer sites, controlled procedures for conducting servicing functions are not necessary.

4.20 Statistical Techniques

86. List four uses of statistical techniques.
87. List seven basic SPC tools.
88. True or False: Statistical techniques are required by the ISO 9000 standards.

6

Relationship of ISO 9000 to TQM, MRP, and JIT

Introduction

Some companies and organizations are managed by reaction. If something breaks, fix it. If you run out of materials, buy more. If things get messy, have a meeting with the appropriate managers. If a situation seems to be out of control, form a committee to study the problem. If customers complain about poor quality, threaten the director of QC. If profits sag, lay off workers. If sales sag, get a new marketing director. This chapter deals with the relationships of ISO 9000 with quality techniques and systems that are more proactive than reactive and that have proven themselves to be very effective.

ISO 9000 plus TQM equals world class quality

The recent pursuit of ISO 9000 registration by many businesses has made some quality professionals question whether Total Quality Management (TQM) and ISO 9000 can coexist. Some traditionalists feel that the ISO activity will somehow detract from the gains made by TQM and cause many mature TQM systems to deteriorate. This chapter shows how TQM and ISO 9000 can coexist and reinforce each other.

TQM

Total quality management (TQM) began after World War II but only gained widespread popularity in the early 1980s. Originally called Total Quality Control (TQC), it represents a philosophy of management style that strives to totally integrate all aspects of quality in a company:

- quality in design and engineering

- quality in production

- quality in marketing and customer relations

- quality in service

- quality in maintenance

- quality in employee relations

- on-going improvement

The Evolution of TQM

Total Quality Management evolved from a "small q" or manufacturing quality focus to a "BIG Q" emphasis on quality for all processes. This developed when enlightened quality professionals realized that controlling only manufacturing processes left a large portion of the quality job undone. Controlling production processes is a good start, but the offering to the customer is made up of the entire package:

- the sales call

- the design process

- the contract negotiation

- the purchase of raw materials

- the inventory process

- the on-time delivery of undamaged product

- customer service

- the training process

Without a quality emphasis on these essential business processes, customer satisfaction frequently suffers.

TQM can be viewed as the antithesis of the "lone ranger" style of management sometimes associated with U.S. business and industry leaders. Although we are fiercely proud of our independence, we are coming to understand that team tactics and integration generally outperform myopic and egotistical individualism. The focal point of TQM is the customer. All management aspects and operations are coordinated to meet all customer expectations and hopefully exceed them. To make this work, customer feedback is essential to making the correct decisions. A systematic approach is required. At one time, quality was the responsibility of one department in an organization. Top management sent orders to the QC (Quality Control) department to "improve quality" and then forgot about it. This only works to a minor extent but will not produce the overall level of quality required to succeed in business today. Presently, the quality system is actually the blueprint for all aspects of a company and guides its day-to-day operations.

Culture Changes

Established manufacturing and service cultures are difficult to motivate toward a new direction. Using ISO 9000 and TQM together can make it happen effectively. ISO develops the structure that insures the processes will be consistently run. TQM promotes continuous improvement to insure that new ways of doing things are pursued. The two systems working together can make significant culture changes accepted in established organizations. TQM makes the gain and the structure of ISO insures that the gain is maintained. Backsliding cannot occur if the new way is documented and institutionalized into the established system. TQM and ISO 9000 working synergistically can facilitate established cultures moving towards world class operations.

TQM is an on-going process, not a mad dash to achieve a goal followed by a period of relaxation. It can be likened to physical fitness. It is unwise to diet and exercise for six months, achieve a goal, and then revert to old habits. True fitness requires a change in lifestyle and an on-going regimen of good nutrition and exercise. TQM requires an organizational culture change that embraces on-going dedication and continuous improvement.

No Conflict

ISO 9000 and TQM are different focuses with the same general objective: quality products or services delivered on time to the customer. ISO uses the traditional approach of a systematic structure that proceduralizes quality activities to insure that "all the i's are dotted and the t's are crossed." TQM views all business processes as affecting quality and, as such, must be controlled and improved to maximize the quality offering to the customer (internal and external).

Some TQM practitioners might say that ISO 9000 is outdated in today's requirement for a continuous improvement mentality. However, ISO 9001's statement of scope explains that the emphasis must be placed on defect prevention. ISO element 4.14 (Corrective Action) mentions "preventing recurrence" which requires making improvements to systems that exhibit problems.

Among other things, TQM requires management involvement, thorough planning, good communications, comprehensive training, and process improvement metrics. ISO 9000 provides the structure for these TQM requirements via:

- ISO element 4.1 Management Responsibility

- ISO element 4.2 Quality System

- ISO element 4.5 Document Control

- ISO element 4.18 Training

- ISO element 4.11 Inspection, Measuring and Test Equipment

The commitment of management to TQM must be nothing short of total and enthusiastic, convincing all employees that quality is a top priority. The quality team includes the CEO, the janitors and everyone in between. It is easy to determine when an organization (any organization) cares about quality. Consider two possible responses to a visitor who meets an employee and asks for assistance:

Response 1: "Don't ask me pal, I don't work in this department."

Response 2: "I don't know, but I will find someone who can help you."

Employees in organizations with successfully adopted TQM realize that quality goes beyond goods and services, and more importantly, each of them knows that they are an important team player.

Relationship of ISO to TQM

The relationship of ISO 9000 to TQM is strong. In fact, one can view ISO as the documented structure for a TQM system. It calls for management commitment and a defined organization that identifies the interrelations among all personnel who affect and control quality in the company. It demands documentation of plans and procedures. It empowers employees to initiate action to prevent product or service nonconformities. It seeks to identify and record all quality problems and provide solutions. It requires the tracking and verification of those solutions.

An ISO 9000 quality system must be documented and include items such as:

- plans as set forth in a quality manual
- the documentation of controls, equipment, skills, and other resources
- the control and approval of revisions to procedures
- identification of required enhancements in measure and control
- the clarification of standards and procedures
- demonstration of compatibility in design, production, and tests

The implementation of a total quality plan is an ambitious project. It requires both the commitment and the participation of the chief executive officer or general manager of the organization. It cannot become a responsibility delegated to someone else. The CEO will have to spend a significant amount of time working on the plan. This sends a clear message about the importance of the process. If this is not done, many employees will ignore the process and some may even resist it. It is well known that human beings tend to distrust change.

As the CEO participates in the process, he or she will be experiencing some of the same problems as the other employees. This will produce empathy and a basis for meaningful communication and progress. The CEO will also have to commit other resources to the process. These will be in the form of funds, space, employee hours, possibly travel and consulting, and so on.

TQM planning is a form of strategic planning in the sense that no part or single issue can be meaningfully discussed in isolation. Strategic issues are strongly interrelated. A strategic planning committee, best chaired by the CEO, is a good way to make sure that issues are discussed in perspective. All segments of the company culture should have representation on this committee. This will secure the proper input and it will send out the message: "We are going to do it with your help."

Once a plan is in hand, it is time to begin implementation. The planning committee should not be discharged from its duties. The meetings may be fewer and shorter, but they must continue. The group's main purpose now is to obtain feedback (from customers too) on how the process is moving. Problems are best discussed and solved by the full group. It may be necessary in some cases to revise the plan based on problems that occur during implementation.

If all goes well, the TQM plan will begin to function, quality will improve, and efficiency and employee attitudes will also improve. Most importantly, the feedback from customers should now be headed in the positive direction. The planning committee becomes a standing management committee at this point and should work constantly to determine how additional gains can be made. The occasional rotation of members is a method often used to safeguard enthusiasm and vitality.

ISO 9000—A Full Service Business System

The ISO 9000 series of standards recognizes the need to emphasize more than just production quality. For example, there are a number of system elements in ISO 9001 that address non-production quality issues:

- 4.1 Management Responsibility
- 4.3 Contract Review
- 4.4 Design Control
- 4.6 Purchasing
- 4.7 Purchaser Supplied Product
- 4.15 Handling, Storage, Packaging and Delivery
- 4.18 Training
- 4.19 Servicing

These elements dovetail nicely with the TQM requirement to go beyond production quality into the full range of issues that affect customer satisfaction.

Complementary Systems

Total Quality Management emphasizes the control and improvement of all processes in a business. This can be best accomplished when a structured quality system is used as a launch pad for continuous improvement. Unfortunately, a vibrant, structured quality system is not always in existence when a TQM effort is initiated. Instead, other situations occur, such the unavailability of a viable structured quality system or an underused quality system. ISO 9000 is the type of structured quality system that can enhance the TQM effort and hold the gains achieved by TQM's continual quest for continuous improvement of all processes.

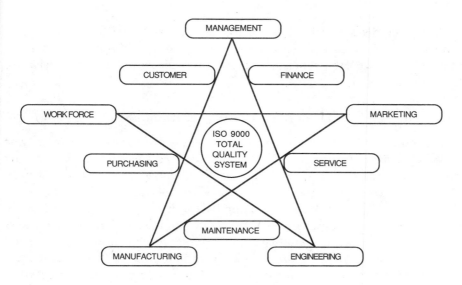

Figure 6.1 ISO 9000 at the center of the TQM structure

Enhancement of SPC

As an example of how TQM can be less effective without an established quality system, take the case of an important TQM tool, Statistical Process Control. We view SPC as an important tool for continuous improvement. Ineffective implementation occurs when the tool is applied in an unstructured environment. Making control charts might be an interesting academic exercise, but the full benefit of identifying and eliminating components of variation will not be realized. A fully functioning quality system that controls documents, quality records, equipment calibrations, audits, training, product design, customer and supplier contracts, and other functions affecting quality must be an integral part of a continuous improvement system.

The advantage of ISO 9000, coupled with the use of SPC in a controlled environment, is that ISO 9000 brings the structure and accountability to SPC implementation. If SPC is incorporated into the ISO 9000 quality system and continuous improvement requirements are documented, then Management Review will check to insure that the continuous improvement metrics are achieved. Corrective Action and Internal Quality Audits also promote continuous improvement efforts related to SPC. The structure required by ISO 9000 becomes a valuable tool for management to insure that objectives established early in a TQM implementation plan are not forgotten when crises in other aspects of the business are encountered.

Document Control Example

A quality engineer for a manufacturer of stamped metal products was responsible for identifying and reducing product variations. The product lines were mature, and over the years, established procedures had been developed. However, these procedures were not tightly controlled even though every operator had been supplied with documentation regarding his or her particular procedures. While trying to identify and reduce variation in the processes, the quality engineer discovered that operators found what they thought were better ways to run the processes. They would then incorporate changes in their procedures but would not tell others what they were doing. This produced product variations that were very easy to see from control charts. Applying the trouble-shooting techniques of statistical process control, the engineer identified the source of the variation, but could not remove the variation until a system for document control was developed and implemented.

By developing a document control system that incorporated strict guidelines for using only controlled documents shared by all, it was possible to get all operators to use the same procedures. The system had the following features:

- All prints, procedures, visual aids, and workmanship standards were considered controlled documents.

- These documents were either produced on controlled paper with red headings or were stamped with a red controlled document stamp.

- The originals of these documents were kept in the documents control room and were only released to predetermined recipients (that list was also on controlled paper).

- If a document change was required, an approval form was initiated and circulated to the appropriate functional groups for authorization.

- Upon approval, the documents were changed and notification of the changes was circulated to the appropriate personnel.

If the document change was required immediately, a temporary manufacturing procedure was initiated during the approval and change process so that the change could be implemented to capture the improvement as quickly as possible. This temporary procedure also required approvals, but it was permissible to submit it in draft form for rapid implementation. The temporary procedure had a defined expiration date and was removed from service by the document control group on the expiration date. The major gains of this controlled document system included:

- The largest component of variation was reduced (operator-to-operator).

- As operators used their creativity to make improvements, a system was in place to institutionalize the improvements and train everyone to do it the better way (often called "best practices").

- The operators were motivated to use creative problem solving since they knew their solutions would be considered.

- Operator attitudes improved due to the recognition of their contributions.

- The documentation system promoted reduced variability on all of the other products in the stamping area.

Total Quality Management emphasizes continuous improvement, but in order to insure that there is a baseline to improve upon and that variation is minimized in all processes (manufacturing and administrative), the documents that define how business is conducted must be controlled. ISO requires documentation control. Although on the surface, ISO and TQM may seem to be totally different quality approaches, there are many similarities. If planned properly, the two systems can work hand-in-hand to deliver unsurpassed value to internal and external customers. Many organizations that already had mature TQM systems have been pleasantly surprised to watch them improve as they worked toward ISO 9000 registration.

MRP

Materials Requirements Planning (MRP) is one way to bring more order to chaos. It is basically an automated (computerized) inventory/production management system intended to prevent shortages, swollen inventories, waste, late deliveries, and confusion. Today, either MRP or JIT (Just in Time) should be considered to systemize and control raw material inventories and work in progress inventories. JIT is covered in the next section of this chapter.

The heart and soul of MRP is the *master production schedule* which is not based on production history but on the best possible data concerning immediate and future needs. The master schedule is *exploded* into purchase orders for materials and work orders for internal operations and processes. As an example, suppose the master schedule calls for the production of 500 electronic instruments. The bill of materials for this product will list all the required materials, parts, and subassemblies whether purchased from the outside or produced internally. If the bill of materials for the electronic instrument shows 4 switches, then 2,000 (500 x 4) will be needed. If current inventory shows 500 on hand and purchasing records show 800 on order, then 700 additional switches will be ordered. The automation is provided by a computer data base and an MRP program which considers the number of production units, the bill of materials, the current materials inventory, and any orders in progress. Lead times on material acquisitions are also a part of the data base so that all materials should be available on time to support the master schedule. Most MRP systems can also handle related materials such as packing materials, shipping containers, manuals, warranty cards, accessories, and so on.

When the MRP system places orders for materials it also schedules all of the necessary internal operations. In the case of the 500 electronic instruments, times are allocated for the production of parts, subassemblies, final assembly, testing, and packaging.

There are several types or levels of MRP systems:

- *Type I*. This is an inventory control system which generates purchase orders and schedules internal operations to meet the timing and quantities specified in the master schedule.

- *Type II*. This is a production and inventory control system that does everything a Type I system does and also checks materials and production capacities. If the capacities are not large enough to meet the master schedule, then the capacities or the master schedule, or both, are changed.

- *Type III*. This is an elaborate system that does everything a Type II system does and also is used to plan and schedule all of the manufacturing resources including personnel, equipment, facilities, finances, and preventive maintenance.

MRP means materials requirements planning in Type I systems. MRP means *manufacturing resource planning* in Type II and Type III systems since they deal with other resources in addition to materials. Today, MRP

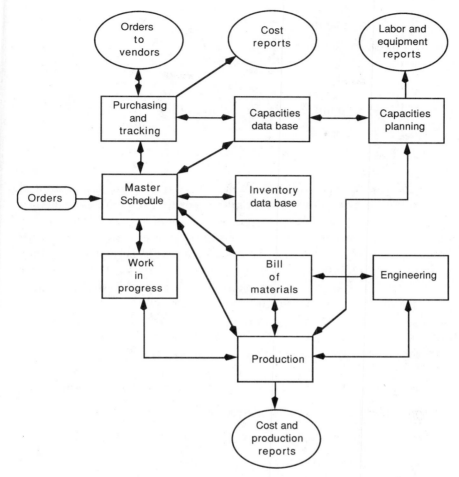

Figure 6.2   Example of an MRP system

systems range from simple programs that run on desktop computers to elaborate, integrated software packages that run on minicomputers or mainframes.

Relationship of MRP and ISO

As with TQM, it can be determined that the generic nature of the ISO standards produce a strong relationship to any system that can affect quality, and MRP certainly falls into that category. For example, ISO 9002 lists the following quality system requirements, some or all of which could overlap an MRP system:

- 4.2 Quality System
- 4.3 Contract Review
- 4.5 Purchasing
- 4.6 Purchaser Supplied Product
- 4.7 Product Identification and Traceability
- 4.8 Process Control
- 4.14 Handling, Storage, Packaging, and Delivery

One of the most significant things to consider when planning the journey to ISO registration is the complete coordination with existing quality systems. This is one of the reasons why team efforts and the participation of top management are both essential for success. If an organization is already effectively using an MRP system, it would be foolish to address an area of the ISO 9000 standards, such as purchasing, without first reviewing exactly how the existing system operates. It may well be that the appropriate documentation for the MRP system will transport directly to the quality manual and satisfy that part of the ISO requirements. Or, it will at least serve as a part of the required documentation.

In the area of purchasing, the ISO standards require:

- that purchased product conforms to specified requirements
- that subcontractors meet all requirements
- records of prior experiences with subcontractors
- purchasing data (type, class, style, grade)
- specifications (drawings, process requirements, inspection details)

Most MRP systems have some or all of this information as a part of their data bases. In fact, it is possible to use some MRP systems to generate reports to satisfy some of the ISO requirements. As the interest in ISO continues to expand, there is little doubt that the designers of MRP systems will integrate more of the ISO requirements into their product lines.

JIT

The ultimate goal of JIT (Just in Time) is to eliminate inventories of raw materials, Work-in-Progress (WIP) inventories, and finished goods inventories. Of course, the ultimate goal is never reached but working to approach it can improve quality, increase efficiency and save money. As a matter of fact, most advocates of JIT concede that it is not possible to entirely eliminate inventories.

Originally, JIT was used to reduce raw material and WIP inventories. Now it has spread to include other areas such as Just in Time Manufacturing (JITM) and Just in Time Training (JITT). The basis of JITM is to produce the exact amount needed and only when it is needed. JITT regards mass training as inefficient. It doles out training in smaller chunks and delivers them as they are actually needed.

JIT can be considered as a total quality subsystem such as SPC. It minimizes waste and saves money. When JIT is working well, there is a significant reduction of warehoused materials. Materials and goods are not purchased or produced until they are needed. This reduces or eliminates handling, spoilage, obsolescence, storage costs, and it reduces damage and loss.

JIT has created a new paradigm for manufacturing. In traditional mass production, parts and raw materials are converted into subassemblies. These pass on to the next stage and are combined with other subassemblies. These move on to the next stage, eventually to the final step, and then out of the production area. It is extremely difficult to balance all of the lines to avoid staging (queues) and the buildup of significant amounts of WIP. When a particular line is not ready to accept input, the output of preceding lines must wait at a staging area which resembles a local warehouse. In some cases, accumulating WIP gets in the way and must be moved out of the production area for a time. The extra handling is expensive and increases the chances of loss and damage. JIT controls the line from the output view. It is often said that the customer controls the line (nothing is built until there is an order for it).

Push Systems and Pull Systems

Conventional mass production is considered a *push system* while JIT is a *pull system*. In a push system, the raw materials are purchased and pushed into the warehouse. Next, they are pushed into the first stage of production, the second stage, and so on. Finally, finished product is pushed into a warehouse where it awaits shipment to the customer. Figure 6-3 shows a traditional system.

In an ideal pull system, customer orders drive the system. For example, an order pulls a unit from the end of the production line. This in turn generates a pull to the preceding production process. This continues until the pull eventually ripples through the system and generates an

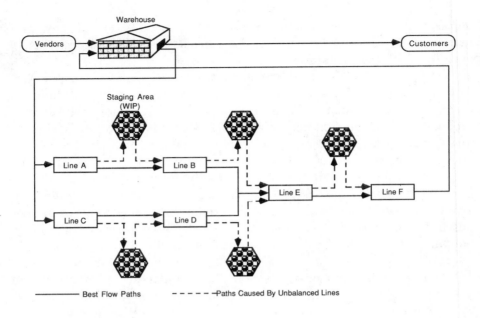

Figure 6.3 Traditional mass production (Push system)

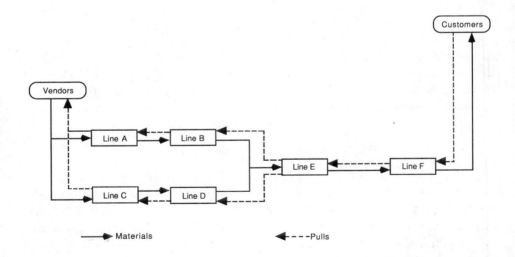

Figure 6.4 Just-in-Time production (Pull system)

order for parts and raw materials. All inventories including WIP are eliminated. Several pulls can be in progress at any given time. In other words, there is no requirement that subsequent pulls must wait until the one in progress ripples all the way through the system.

The Japanese use the word *Kanban* (pronounced kahn-bahn) to identify one methodology used in some of their JIT systems. Kanban is the Japanese word for "card." Cards are used to signal the need for more material. Actually, there are one card systems, two card systems, and three card systems. The one card systems are the most prevalent and these use withdrawal cards. Because the cards are often the most visible part of Japanese JIT, some people believe that Kanban and JIT are synonymous terms, but they are not. JIT systems can operate without the use of any cards.

For an example of a simple withdrawal card system, refer to Figure 6-4. Suppose a tray of parts is emptied at Line E. The tray's withdrawal card identifies Line B as the preceding process. The empty tray and its associated withdrawal card are moved to Line B. The card identifies Line E as needing the parts and it also identifies the parts. Line B fills the order and the tray and card are moved back to Line E. When the trays are full, production stops. Also, the trays and the cards are visible indicators of work flow. Figure 6-5 shows an example of a withdrawal Kanban.

The problems related to the inventories and WIP associated with traditional mass production include:

- consumes valuable space (both warehouse and factory)

- money is "tied up" (not available for other purposes)

- unfavorable tax situation

- requires control, protection and handling (these do not add value)

- spoilage (oxidation, contamination, deterioration, and so forth)

- damage and loss

- obsolescence

- confusion and mistakes (parts mix-ups, items misplaced, and so on)

- wasted motion and delays

Fat Hides Problems

Some of the often cited strengths of traditional mass production, with its accumulations of material inventory, product inventory and work-in-

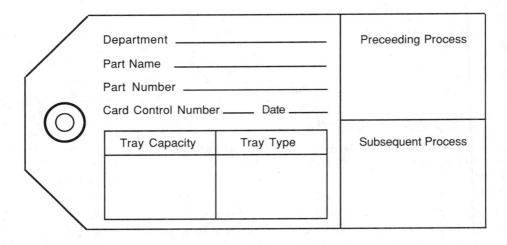

Figure 6.5 Withdrawal Kanban

progress, is that the production of final product is buffered from problems such as breakdowns, work stoppage on one line, failure of suppliers to deliver on time, and so on. Supporters claim that without these buffers, it will be impossible to ship to the customers in a timely fashion. Although excess inventory does provide a buffer, the question must be asked: "Is this method of doing business costing too much?" Remember that all forms of inventory incur costs and other problems. A more insidious problem with traditional mass production is that it tends to mask other costly problems and tend to reduce quality.

Consider the analogy of an over-powered vehicle. Such a vehicle cannot get the job done any faster because there are speed limits and safety considerations. Since the vehicle is over-powered, the driver will not notice minor problems. A faulty spark plug, a clogged injector, and a vacuum leak all go unnoticed because the vehicle can still achieve the speed limit. Although the vehicle gets the basic job done (it gets there) it costs a lot more to operate due to reduced fuel efficiency. It also contributes more than its share of pollution to the environment.

Now consider a vehicle with marginal power. The driver is likely to notice any malfunction because the effect on performance is greater. The driver is forced to attend to problems as they occur and the vehicle is maintained at a high level of performance and efficiency. In fact, the driver is more likely to engage in a schedule of preventive maintenance since she or he has learned that maintenance is critical to performance. This is what

JIT can accomplish in an organization. It is a "lean and mean" design that should be constantly nurtured to yield reliability, quality, and efficiency. JIT demands:

- management commitment to total quality organization in all departments
- concise and documented arrangements with reliable suppliers
- reliable lines with little or no work stoppage
- elimination of unnecessary steps
- reduced cycle times (efficient lines)
- quality (all or nearly all work must be fit to move on)
- preventive maintenance (no fat to buffer breakdowns and poor quality)
- continual elimination of problems (sometimes before they happen)
- on-going improvements

JIT can be tough to promote to those who are comfortable with traditional mass production. It is well known that a stopped line with idle workers is about the worst thing that can happen in a traditional setting. Managers are happy when everyone is busy and the output is high. Line supervisors know this and keep the lines running even when there are known problems. Quick fixes, work-arounds, and unauthorized adjustments are the orders of the day. Quality drops but nobody notices (at least not at that time).

JIT works differently. Process B orders a sub-assembly from process A. It arrives but it has a problem. There is no choice but to shut down process B and this effectively stops process A since no more pulls can be sent to A. Now, the A and B workers are all standing around. This commands a lot of attention and the problem is resolved as quickly as possible. Ideally, the A and B workers and the production supervisors will all contribute their ideas and work together to solve the problem. With JIT, process A cannot crank out a quantity of below standard sub-assemblies that will have to be reworked later or scrapped.

Forming Partnerships With Suppliers

The word *partnership* is positive because it implies a symbiotic relationship where each participant's success is strongly linked to the well-being of the other participant(s). Adversarial relationships will not lead to the long term success of vendors or buyers. The vendors must understand and be willing to comply with JIT. Buyers should not be forced to maintain inventories. It is not unusual for several deliveries of the same materials to occur during one shift in a JIT factory. JIT can be enhanced by electronic data interchange between vendors and buyers. Partnerships with vendors and subcontractors must be developed and nurtured in order to meet the real spirit of TQM and to realize the best possible gains from MRP or JIT systems.

ISO and JIT

ISO Article 4.1.1 requires management to define and document its policy and objectives for, and commitment to, quality. A JIT system enhances quality beyond what a QC department can ever hope to accomplish operating alone. In JIT, each pulled part or subassembly must be acceptable for the manufacturing process to continue. Article 4.1.1 goes on to state that the quality policy must be understood, implemented, and maintained at all levels in the organization. In ISO Article 4.1.2.1, personnel who manage, perform, and verify work affecting quality are identified as needing the freedom and authority to initiate action to prevent the occurrence of product nonconformity and to initiate, recommend, or provide solutions. It further empowers them to verify the implementation of solutions and to control further processing of non-conforming product until the deficiency has been corrected. *This is precisely how JIT is supposed to operate.* Obviously, ISO 9000 and JIT reinforce each other and there is no reason why one would preclude or replace the other.

Chapter Review Questions

1. Are there any major conflicts between ISO 9000 and TQM? Why?
2. What is the difference between TQM and TQC? Which came first?
3. List the major elements of a TQM system.
4. Cite some specific examples of how TQM uses a global view of an organization's structure.
5. Explain how TQM is a "culture change" as opposed to just another company crash program.
6. When TQM is in place, which company employees are members of the quality team?
7. Is it true that ISO can be viewed as a way to structure and document any TQM system? Why?

8. Why does TQM demand a sincere commitment from top management?
9. How is it that quality tools such as SPC are sometimes little more than an exercise or just something to show to customers?
10. Why should tools such as SPC be a part of the total quality system?
11. Describe some of the techniques that can be used to control documents.
12. Can a controlled document system reduce process variations? How?
13. Explain the term *best practices*.
14. Discuss some of the ways that MRP can improve both efficiency and quality.
15. Explain how materials requirement planning can evolve to manufacturing resource planning.
16. List the ISO requirements in the area of purchasing.
17. What is the basic concept of JIT?
18. How can an effective JIT system improve quality?
19. How can excess WIP endanger quality?
20. Contrast and compare push systems with pull systems.
21. What is a Kanban?
22. Discuss how fat or excess capacity can hide problems.
23. Traditional mass production tends to hide defects while JIT tends to correct them in short order. Explain why.
24. Why is it crucial to form partnerships with suppliers when JIT is used?
25. Explain how ISO and JIT can reinforce each other.

7

ISO 9000-3: QUALITY IN SOFTWARE

INTRODUCTION

The software sector of American industry is rapidly becoming one of the most productive business domains in the world. Today, computers and computer software permeate our culture. Many people depend on software as an integral part of their daily existence. In addition to using software for such tasks as word processing, communications, and accounting, embedded software is used every day in other products. For example, smart thermostats use software to control building temperatures. Software also triggers the light in many motor vehicles as a warning when it is time to visit a mechanic. Software lurks behind the programmable keypad on microwave ovens. Any company developing and producing software as its primary product, or as an embedded system within another product, should be concerned about the quality of its software development and maintenance process.

EMERGING STANDARDS

Quality has become a significant concern within the software industry. As people begin to depend more and more on software, increasing demands are being placed on software quality. Many software companies are beginning to look at ISO 9000 as a model for a quality management system. However, the direct application of standards developed mainly for a manufacturing environment to a software development environment is questionable, at best. Like manufacturing, the software development process is based on the five stages of the product life cycle: contract, design, production, installation, and servicing, but software development involves an entirely different set of concerns and constraints.

In a manufacturing environment, the majority of quality control processes are directed at the production stage of the product life cycle. It is during this stage that the majority of product non-conformities occur. The challenge lies in creating consistent and effective production line procedures, which generally lead to increased quality in the finished product. On the other hand, in a software development environment, the design and servicing stages of the product life cycle require the most attention to quality. Design is the most fundamental aspect of developing and producing a piece of software. Over 50 percent of the bugs (defects) in a software product are due to poor design. Servicing is also an important stage of the software product life cycle. In the manufacturing industry, the supplier generally does not need to send the customer add-ons to increase functionality. Typically, the supplier simply produces and markets a new and improved product.

Software developers must design and produce bug fixes and patches, and often, modifications and enhancements to the product. Software must be maintained not only to meet customer's changing requirements, but the requirements of new hardware and operating systems. If these modifications are not correctly made, and do not fit into the original design of the system, the software product will begin to erode and eventually, require complete redesign. Because of these differences, special consideration must be given to the issue of quality in software development.

In the past, the issue of quality in software was addressed in a simple manner—if only a small number of bugs were identified during testing, the product was deemed a quality piece of software. However, as software began to increase in complexity and project scope, it became more and more difficult to find bugs, and to eliminate them without inadvertently altering other important functionality in the software. It quickly became evident that, in order to build quality into software, the developer's management would need to retain strict control over the entire software development process, not just the testing stage.

This is why, in 1991, the International Standards Organization released ISO 9000-3, guidelines for the application of ISO 9001 to the development, supply, and maintenance of software. Similar to the other standards in the series, ISO 9000-3 provides a generic set of guidelines for implementing a quality assurance system. The ISO 9000-3 document provides recommendations, rather than requirements, for implementing a quality assurance system in a software environment. The standard does not dictate what types of programming languages, design methodology, or software development conventions should be used within the development environment. It provides guidelines for insuring that all activities relating to the entire software product life cycle are controlled and measured for effectiveness.

ISO 9000-3 was written primarily for two-party contractual situations, where a supplier must develop a software product according to the requirements (as stated in a contract) of a specific customer. The standard can easily be adapted to other situations, such as software developed to meet a market need (based on marketing studies and analysis) or to a software product developed for internal use only. In cases such as these, the marketing department or internal users would become the purchaser or customer, and the contract would evolve from the results of marketing surveys and analysis, or lists of requirements generated by the internal users. In this chapter, the word *supplier* refers to the software developer/provider.

Currently, there is no such thing as registration to ISO 9000-3. Companies seeking a software registration must register under ISO 9001, using the 9000-3 document as a guide. While it is perfectly feasible to develop a quality system using 9000-3, it is important to note that any nonconformities cited during the audit process will be based on 9001. Therefore, it is a good idea to use the 9000-3 document as an *interpretation* of 9001, with care taken to address the elements of 9001 in the quality system. Another registration option for software companies is the TickIT program.

The TickIT Program

TickIT is a British certification body which addresses quality concerns in the software industry. TickIT stands for "Check IT (Information Technology)". TickIT registration is a separate process from ISO registration, but uses ISO 9001 as a model for quality systems and 9000-3 for guidance. During the audit process, the TickIT auditor uses the *TickIT Guide,* an audit guide based on ISO 9000-3. TickIT auditors must have software-specific education and/or industry experience, and pass an oral interview about selected software engineering topics. For more information about TickIT and TickIT registration, contact the DISC TickIT office:

DISC TickIT Office
2 Park Street
London, WIA 2BS
Voice: 011-44-71-383-4501
FAX: 011-44-71-383-4771

The TickIT registration scheme has not yet gained international acceptance.

SQSR

The SQSR (Software Quality System Registration) committee is currently addressing the possibility of an ISO software-specific registration process in the United States. If accepted, the SQSR program will be based on the same fundamentals as the TickIT program. Registrars will be certified by the RAB under the SQSR scheme. Auditors will be expected to have industry experience in software and quality auditing and to pass an oral interview. Auditors will also be expected to complete formal software engineering training. The SQSR program was presented formally to the RAB in 1993. Currently, the issue is still being debated.

QUALITY SYSTEM FRAMEWORK

ISO 9000-3 shares a few basic elements, mainly those dealing with the framework for a quality system with the 9001 standard. This chapter will not go into detail about the elements taken directly from the 9001 standard. For more information on these elements, please refer to Chapter 5.

4.1 MANAGEMENT RESPONSIBILITY

The elements regarding management responsibility are exactly the same as outlined in ISO 9001, Section 4.1, with one exception: the responsibilities of the purchaser's management must be clearly identified and defined.

Software is generally developed to meet a customer's functional and operational requirements. These requirements are often complex, and can easily be misinterpreted or misunderstood. Therefore, successful communication between the supplier and purchaser throughout the software development process is critical. Requirement specifications must be reviewed, design iterations must be reviewed and tested, requirements must be updated, and a final acceptance test must be administered. In most cases, the development is also dependent on information or data to be provided by the purchaser. It is the joint responsibility of the supplier and purchaser to insure that the supplier receives all necessary information on time, and that this information is not misinterpreted.

Purchaser's Management Responsibility

The purchaser should assign a representative to function as the key go-between for all contractual matters with the supplier. This representative will need an adequate level of administrative authority to deal with contractual matters, such as approving and changing the requirements specifications and defining final acceptance criteria, as well as the technical knowledge to clearly answer questions the supplier may have. It is also important for the representative to have some interaction with end-users of the software. Many times, software is well developed from a functional perspective, and less developed from an operator's perspective.

The selected representative should also be responsible for dealing with any delivered software that does not meet the purchaser's expectations. The establishment of a successful working relationship early in the development life cycle can have enormous benefits for both the supplier and the purchaser. Software development does not occur in a vacuum. It is an iterative process of design, review and testing, and it is very important to involve the purchaser in these activities. For example, imagine a finished application is delivered and the customer says, "This software doesn't do what we want! We can't use it!" The representative failed to communicate effectively with the customer and the damage is done.

The supplier should produce a document (a contract) which states the responsibilities of the purchaser. This document should identify the purchaser's management representative, and all associated responsibilities. In addition, the supplier must maintain records of any changes or clarifications made to the contract and/or requirement specifications.

Joint Reviews

In order to further enhance professional relationships and reduce ambiguities in the contract and specifications, joint reviews between supplier and purchaser must be scheduled at regular intervals. Both parties should meet to insure that the software is being developed to meet the actual needs of the customer. It is a good idea for the customer to visit the supplier during one or more stages of development (depending on the length of the project), to work with the software while it is still in its development stages. A working relationship limited to memos, faxes, and telephone conversations will ultimately lead to misunderstanding customer needs, or worse, overlooking those needs altogether. In addition, all verification and acceptance test results should be made available to the customer. These test results should be retained as software development quality records

The supplier must document the procedures and schedule for joint reviews. This document should contain information such as review location, review participants, frequency of reviews, issues addressed, and so forth. The supplier will also be responsible for keeping records and minutes of all reviews. Any action items resulting from the review must be explicitly noted.

4.2 QUALITY SYSTEM

The supplier must establish and maintain a documented quality system. This quality system should be incorporated into every aspect of the software development life cycle, and should provide management with sufficient control over the development as it progresses. The quality system should focus on proactive measures, rather than reactive measures. "Problem prevention should be emphasized rather than depending on correction after occurrence." It is the supplier's responsibility to insure that the quality system is implemented, but the standard does not provide specific procedures for accomplishing this. The supplier is expected to implement the quality system in a way that will meet the needs and constraints of the individual organization.

Quality System Documentation

The supplier should prepare a *Software Quality Manual* that covers all the elements of ISO 9000-3. The manual should serve internally as a formal procedures manual, and externally as evidence for customers interested in the software development process as subject to quality control by management. The manual should cover all aspects of software development at the organization, including:

- *organizational overview*—description of the product to be delivered and the overall structure of the organization

- *responsibilities*—who is responsible for which activities and how they all interrelate

- *tools*—all software development tools that are used in development, which might include such things as third-party compilers, bug tracking systems, and configuration management software

- *methodologies*—design techniques and life cycle models used. (References to other documents should be included in this section.)

- *standards*—programming languages used, internal source code format requirements, user interface guidelines, and so forth. (This section should also include references to other documents, such as programming guides and corporate guidelines.)

- *other relevant items*—documentation standards, test procedures, contract review procedures, training and qualification, records management, audit procedures, and so on

Basically, the manual should reference all quality procedures and activities associated with each step of the development process—from design specification writing to preparation of user documentation, to the final delivery of the software product. Management should work closely with employees to insure that all procedures are performed, as documented, within the organization. The quality manual should be clearly written and readily accessible to anyone in the organization.

The quality manual should be an overview of the quality assurance system, and it should reference associated documents, as needed. For example, procedures may be outlined in separate documents and referenced in the quality manual. Procedures should be defined in simple terms and written at a level that can be easily understood by all employees. The quality manual should be reviewed and updated on a regular basis, so that it continues to reflect the structure of the organization. The supplier should keep records which reflect the issue, review, and approval of the quality manual, and any changes must be approved and recorded.

Quality Plan

For each software development project, the supplier should document a quality plan. Every software development project is unique in its own way. The quality plan should outline any areas in which the project may diverge from the procedures and standards outlined in the quality manual. Any special circumstances or requirements for a particular project should be documented in the quality plan. Similarly, any special customer requests concerning quality should be documented at this time. For example, a customer may require some additional testing or verification procedures beyond those outlined in the quality manual. The quality plan should also outline how quality activities, as documented in the quality manual, will be applied to a particular software project. Everyone involved in the project, including the customer, should review and approve the quality plan before the project is actually started. Records of all reviews and approval status must be documented in quality records.

Internal Quality System Audits

It is critical that the supplier insure that the documented quality system is actually being used. This element addresses the issue of internal audits. This element is taken directly from ISO 9001, Section 4.17. Generally, the auditors must have a software background so that they are in a position to insure that the procedures are actually being followed. Please refer to Chapter 5 for more information on designing and implementing internal quality audits.

Corrective Action

The supplier is responsible for correcting any deficiencies found at any time, including deficiencies found during an internal audit. This element is also taken directly from ISO 9001, Section 4.14. Refer to Chapter 5 for more information on establishing and documenting procedures for corrective action.

QUALITY SYSTEM LIFE-CYCLE ACTIVITIES

All software development projects should be based on some type of life-cycle model. Life-cycle models may vary somewhat in basic concepts, but most involve five basic development phases: system level planning, analysis, design, testing, and maintenance. The different models usually vary according to the implementation sequence of the phases, the number or iterations of each phase, and so forth. ISO 9000-3 does not require the supplier to use any particular life-cycle model. The guidelines suggest that all quality activities within the organization be planned and implemented according to the supplier's life-cycle model. Regardless of the specific model used, the supplier should document that sufficient controls and measurements are being employed during each stage of the development cycle. The supplier should also document the development processes followed in the organization, in addition to a description of the life-cycle models used.

5.2 CONTRACT REVIEW

The *contract* is a document between the supplier and the purchaser outlining the functionality in terms of performance requirements. The contract will be delivered in the completed software product, generally within a specific time frame. For companies not operating on a two-party basis, the contract may be a document based on meetings between engineers and marketing representatives or other employees. In two party arrangements, a product is specifically designed for one customer.

The supplier should have a documented procedure for contract reviews. The procedures should include a list of persons, by job title, who will participate in reviews, as well as step-by-step procedures for conducting each review. A contract review should insure that the contract meets the following minimum requirements:

- clearly and unambiguously defines all software requirements

- clearly identifies any risks or contingencies (for example, if coding of a certain function cannot begin until data is received from the customer, this should be clearly outlined in the contract)

- protects any information that is proprietary to the supplier or purchaser

- does not promise anything that the supplier does not have the capability to deliver—(for example, if the supplier does not have any programmers trained in C++, the contract should not promise to develop a C++ program in six months!)

- clearly identifies the supplier's responsibility for any work that will be performed by a subcontractor

- uses terminology that is agreed upon by both parties

- does not leave any unresolved issues in the proposal

- does not require anything from any party that they do not have the capability to deliver

- defines a procedure for addressing problems and requests for enhancements

Minutes and records of all decisions made during contract review sessions must be documented and stored with quality records.

Contract Items on Quality

It is a good idea to address any potential quality concerns directly in the contract. Quality concerns may include such things as:

- final acceptance of the delivered product

- handling problems found in the delivered product

- processing changes to requirements specifications

- responsibilities of the purchaser and supplier during the length of the business partnership

- standards and procedures to be followed

- requirements and restrictions for replication of the final product

- servicing and maintenance after final acceptance

If procedures for handling these issues are adequately covered in the contract, a great deal of confusion can be prevented during the actual project implementation. It is very common for the purchaser's requirement specifications to change during the life of the project. As the purchaser works with development versions of the program, or changes evolve in the customers working environment, enhancements and modifications to the software will inevitably become necessary.

Some common questions and concerns that evolve during development may be: How will the supplier handle these requests? It is a good idea to please the customer, but not to fall behind in the project schedule. What about the impact of additional enhancements on the cost of the software? How can this additional cost be estimated? If these concerns are discussed between the purchaser and supplier in the earliest stages of the project, policies can be added to the contract. If these concerns are addressed properly, a clear understanding will result and the business relationship will not be jeopardized when the supplier has to say no to a request for an enhancement, or when the supplier decides that the enhancement will cost additional money.

5.3 PURCHASER'S REQUIREMENTS SPECIFICATION

The development of the purchaser's requirements specification is one of the most important aspects of the entire software development process. This document is a functional description of the software product to be delivered. All system design will be based on this document. Without a clear set of functional requirements, it is impossible to develop software that will meet the needs of a customer. These requirements also play an important role in the final approval of the completed software, as the acceptance criteria for the final system-level acceptance test is generally taken from the requirements specifications.

According to the 9000-3 document, "in order to proceed with software development, the supplier should have a complete, unambiguous set of functional requirements. In addition, these requirements should include all aspects necessary to satisfy the purchaser's need. These may include, but are not limited to, the following: performance, safety, reliability, security, and privacy. These requirements should be stated precisely enough so as to allow validation during final product acceptance." A purchaser might be naive about how specific these requirements should be and how important they are. Guidelines or questions should be used to help customers think logically and clearly about their real needs.

If there is to be any type of interface to be developed between the software product and any other software product or hardware product, it should also be clearly outlined in the requirements specification. Generally, the requirements specification is organized as follows:

- *system overview*—defines the scope and basic objectives of the system

- *major functions*—defines all functions and subfunctions to be included in the system

- *installation site*—identifies users and locations and all associated hardware and software

- *design constraints*—outlines any significant constraints assumed in the design of the system (for example, the maximum intended users, data limitations, and so forth)

- *operational constraints*—outlines any operational constraints assumed in the operation of the system (for example, dependence on other systems for input, system overload, and so forth)

The requirements should be developed through cooperation between the customer and the supplier. Both parties should approve the requirements specification before the development activities begin. The supplier must document all procedures, outlining how the requirements will be

generated, reviewed, approved, and changed or updated. Each time a requirements specification is developed for a customer, it should be done
according to these procedures. The supplier must have a requirements
specification for each software development project. Each specification must
be carefully reviewed and approved by both the supplier and customer.
Minutes must be kept of all requirements review sessions, and records of
approval and change processing must also be retained.

The requirements specification is subject to documentation control
and configuration management. It is absolutely critical that the requirements specification remain, at all times, a controlled document. Everyone
in the organization must be made aware of any changes to the requirements specification. See the later sections on configuration management
and documentation control for more information.

Confusion evolves when numerous versions of the requirements specification are circulating within an organization. Even worse, imagine what
can happen when changes are requested by the customer, but not added to
the official document at the supplier's end. Take XYZ Software, Inc. as an
example:

> A rather complex application was being developed to meet a set
> of customer specifications. The organization followed a simple
> life cycle model: design, code, test, and document. Generally, the
> programmers would write the code for a specific unit or function,
> recompile the development version of the program, then inform
> the documentation and testing departments about the new func-
> tionality. The new functionality would then be documented, based
> on the user interface and interviews with the programmer. The
> documentation department would then forward the draft to the
> testing department, where the new functionality and associated
> documentation would be tested.

The above example bogged down because the requirements specifications were outdated. Specifications had been documented at the onset of the
project, but the changes and modifications were conveyed to the programmers by word of mouth, memos, and meetings. The original requirements
specification was never updated. What inevitably happened was that the
programmers developed coding for a specific function, according to the original requirements specification, when the requirements in that area had in
fact been changed. This resulted in lost hours, because the programmers had
to rewrite sections of the code, the documentation department had to change

the manual, and the testing department had to retest the unit. XYZ Software, Inc. is currently three years behind in their contract.

In cases where there is no contract, this document should simply be a software requirements specification, which outlines the requirements for the system that is to be developed.

Mutual Cooperation

In some situations, the purchaser will provide the complete requirements specification. However, in most cases, it is the responsibility of the supplier to generate this document. Generally, the supplier develops the requirements based on the contract. The supplier should not rely on the contract as the only basis for the requirements specifications. The supplier should work closely with the purchaser to develop requirements that adequately reflect the needs of the users. Once both parties sign off with approval on the specification, it becomes the working contract and the basis for the system design.

Both the purchaser and the supplier should assign a representative who will be responsible for all activities related to the development and maintenance of specifications. Methods for approving and implementing changes to the requirements should be agreed upon. The supplier should also insure that the purchaser's representative has consulted with, and adequately represents, the actual end users of the software.

5.4 DEVELOPMENT PLANNING

Before starting any project, it is always a good idea to have some type of plan to follow. In order to produce a quality software product, the supplier must carefully plan the development process, and have a means for insuring that the development plan is being followed. The plans should be subject to preapproval before the project is started, and should be updated as the project progresses.

Development Plan

The development plan is project specific and is a product of the quality system. It must be written in accordance with the quality manual. It follows a protocol to insure that no key items are overlooked. Basically, the development plan defines the project, describes the project resources, details the development phases of the project, outlines the project schedule, and identifies

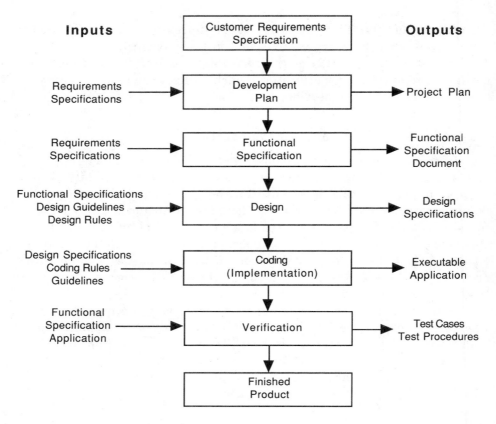

Figure 7.1 Common development phases

all related plans. The plan should also highlight dependencies between tasks and define critical paths. The supplier should develop one plan for each software project. The development plan should be reviewed by all parties involved with, and affected by, the development process.

According to 9000-3, the development plan "should define a disciplined process or methodology for transforming the purchaser's requirements specification into a software product." Generally, this involves separating the work into a series of development phases. These phases usually require specific inputs and generate required outputs. It is also a good idea to plan for some type of verification procedures during each stage in the development cycle. This allows the supplier to prevent nonconformities early in development. The plan should also anticipate and analyze any potential problems that might occur during development.

Generally, the outputs from one phase become the inputs for the next phase. Figure 7.2 provides an example of some common development phases.

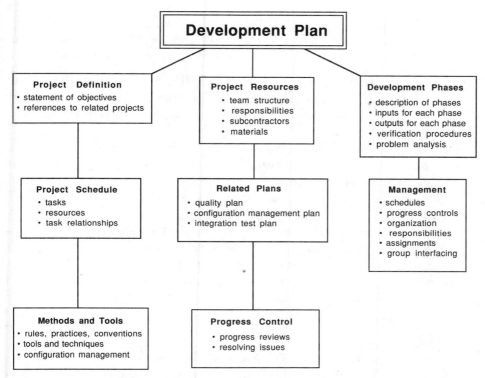

Figure 7.2   Project development planning

Please note that this is a general example. Your company may include many more steps in its development process, or it may have less.

The development plan should also outline how the project will be managed. This should include a complete schedule, including dates for internal releases and deliveries. Progress review schedules should also be documented in the plan. This will allow the supplier to insure that the development is proceeding according to the development plan. Minutes and the results of these reviews must be documented in quality records.

Organizational responsibilities must be documented in the development plan as well. One section of the document should include job titles, job descriptions, and their associated responsibilities. The following are offered as samples:

- Project Manager—The project manager is responsible for preparation and implementation of the project plan. The project manager also has the responsibility to assign tasks, review and approve all project activities, and most importantly, track the plan and ensure that the development is on schedule.

- Technical Responsible—This individual is responsible for implementing the technical aspects of the project plan. The technical responsible executes all tasks as assigned by the project manager.

- Tester—A tester is responsible for testing all aspects of the system under development. The tester is expected to generate and review documented test cases and keep records of all testing activities.

The supplier documents how employees work together on the project. This information covers all technical and information exchanges, as well as basic organizational structures.

The development plan also defines any methods used by the organization to improve and validate the development process. For example, the supplier should identify or make reference to other documents that identify the observed development methodologies and conventions. Any tools or special techniques used during the development cycle should also be identified, for example, third-party development tools (software), configuration management tools, and so forth.

Progress Control

The development plan includes a schedule for progress reviews. The purpose of these reviews is to insure that the development plan is being properly implemented. Any unresolved development issues are addressed at this time, for example, subcontracted work that has not yet been contracted out, unassigned tasks, team members who are not aware of their responsibilities, and so forth. The minutes for these reviews must be documented as quality records.

Input to Development Phases

Development phases must be defined in terms of required inputs and outputs. For example, the program must be at a certain level before the user documentation can be started. The writers must be able to understand how the program will work, they must be able to do screen captures, and so forth. These types of requirements must be defined in the development plan. This will clearly separate the development stages and eliminate confusion.

Output From Development Phases

The outputs from each development phase must be defined and documented as well. There should be some type of acceptance criteria for determining whether or not the project is ready to proceed to the next phase of development. This criteria should be outlined in the development plan. The outputs of each development phase should be systematically verified.

Verification of Each Phase

The development plan includes a description of all verification activities and how they will be implemented. Verification determines whether or not, based on the outputs of one phase, the project is ready to move on to the next phase. According to 9000-3, verification can be attained through development reviews at appropriate points in the development phases; comparing a new design with a proven design, if similar; and undertaking tests and demonstrations. The project proceeds only after each phase is verified.

The personnel responsible for verification can vary. Software engineers will often verify their own work at one level, then move the product to a separate testing group for verification at a higher level.

5.5 QUALITY PLANNING

The supplier must prepare a quality plan to accompany the development plan. The quality plan addresses all issues covered in the development plan. It must be reviewed and approved by all functional groups involved in the project. According to 9000-3, the quality plan can be "an independent document (entitled quality plan), a part of another document, or composed of several documents, including the development plan." The supplier must have a documented quality plan for each project.

Quality Plan Content

The object of the quality plan is to address all quality concerns and objectives for the entire project. This includes verification of the inputs and outputs of all development phases, criteria for inputs and outputs, and details of validation (schedules, activities, resources, and approval authorities). The quality plan should be written from a verification and validation standpoint—you should describe how you will verify the quality of the product in each stage of its development. Responsibility for all quality-related activities must also be defined in the quality plan. Who is responsible for reviews of the development progress? Who will execute validation tests? Who will do configuration control? Who will insure that all corrective action is taken?

The quality plan is subject to formal reviews and approval. It should be updated as the project progresses. Any changes to the quality plan must be documented in quality records. The supplier must document its procedures for producing a quality plan. These procedures should then be used to produce a quality plan, subject to review, approval, and updating, for each development project.

5.6 DESIGN AND IMPLEMENTATION

The design and implementation stages of the software life cycle are where the actual software product begins to emerge. The design process should focus on the data structure and architecture of the software, the procedural details, and the interface characteristics. Written requirements specifications are transformed into fully functional pieces of software. While it may sound magical, design is a very complicated process especially subject to errors. There are many places where a design can go wrong, and some common design errors may fall under categories such as these: interface design, error checking, program logic, performance limitations, and so forth. Once a design has been completed, reviewed and approved, errors can occur during the implementation, or coding, of the system. After all, to err is human. This is why it is so important to control the design and implementation of a software product—to insure that quality is a foremost concern for everyone involved.

Design

The design phase encompasses a number of complex constraints and considerations. Every software company has its own design methodologies and conventions and these may vary from one product to another. For example, the same set of design conventions would not be adhered to when developing a database application as opposed to a non-database application. In addition, the supplier must take into account its own past experiences. What has worked in this type of situation in the past? What has not worked? The supplier should use past lessons to avoid problems in the current project.

Finally, a product should be designed with regard to its future. It should be easy to test, maintain and use, and designers must always keep these factors in mind. The supplier is responsible for documenting all steps involved in designing a product, including design reviews. The design process involves everything that turns a customer requirements specification into a design capable of being implemented. It is very important to control the design process, and documented procedures must be maintained for controlling the following:

- *Inputs*—The design input documentation, mainly the customer requirements specification, should be subject to document control and configuration management.

- *Developers methods and procedures*—You should have a developer's standard procedures manual. This manual discusses methodology, design techniques, documentation tools, testing methods and comments, design document templates, coding standards, and so on.

- *Design feedback*—You should have a system for collecting and using design feedback—from previous development efforts, as well as review of new designs. You should establish a review and approval system for all designs.

All design procedures must be reviewed and approved. A written software design specification must be available for each functional unit of the system. This document is usually written by one or more developers, and is subject to review and approval by others on the development team. Records of all design reviews must be kept, as well as documented procedures for approving changes to the design specification. All changes proceed through the proper channels for approval and estimation of the impacts, and records of all changes must be maintained in quality records. Once approved, design documents should be placed under configuration management.

Decisions made during the design process are critical. A poor design often means that the software product will not be very easy to use or maintain. The design phase provides a representation of the software for quality assessment. A document may be incorporated into the quality system documentation that discusses the elements of a good design.

Implementation

Implementation involves translating a design into an executable software program. This process must also be controlled and disciplined. According to 9000-3, the supplier must address the following aspects in its documented implementation procedures:

- *Rules*—The supplier should have documents which outline basic programming rules and implementation methodologies. For each programming language used, there should be a programmer's guide. In addition, the supplier should discuss consistent naming conventions and commenting rules. Every software company should have some type of internal source code documentation. These rules allow an organization to produce consistent, familiar, and readable source code, and should be followed at all times. If they are not followed, a rationale should be documented.

- *Methods and tools*—All software tools used in development can affect the quality of the software product. These tools must be placed under strict control. All methodologies and tools used must meet the requirements outlined in the specifications.

Where appropriate, standards should be referenced and used, like the American National Standards Institute who publishes standards for many programming languages. The supplier must document and describe all implementation processes which are then reviewed and approved by the organization. All source code must also be tested, reviewed and approved. Records of source code reviews and approval must be maintained in quality documentation.

Reviews

Reviews can occur at various levels of the software development life cycle. According to the 9000-3 document, the supplier must carry out reviews to insure that the requirements are met and the above methods (design and implementation) are correctly carried out. Reviews serve to detect and remove defects early in the development cycle. Reviews also give other members of the development team a chance to voice their own opinions and suggest improvements. The supplier must conduct reviews at both the design and the implementation level. These reviews generally include informal and formal design reviews and code inspections or walk-throughs.

The supplier should conduct both high and low level design reviews. It should be noted that approximately 50 to 65 percent of all errors are introduced to programs during the design phase of the development life cycle. A design review should detect errors in functionality and program logic. Design reviews should also insure that the design meets all the requirements stated in the specification. In addition, the review should check that the design is in line with all methodologies and conventions observed in the organization.

The supplier should also conduct low-level code inspections, or walk-throughs. It is critical for the supplier to insure all coding rules and standards are being followed. An effective code review process can significantly reduce the number of errors in the code. In order to insure that every code review is effective and thorough, the supplier should design a checklist for code inspections (see the sample form in Appendix D). The checklist should include various aspects of the code to check: misspellings and typographical error, adherence to design requirements, program language conventions, and so on. Well-designed review forms can provide metrics for monitoring the effectiveness of the review process. For example, information such as the amount of information reviewed or inspected in a given time period, the number and severity of errors detected during the review, the estimated preparation time for the review, and the number of people participating in the review all provide valuable data for evaluating the effectiveness of the review process. Records of all code inspections must be kept in the quality documentation.

The supplier must have documented procedures for conducting reviews, including a list of the types of features to be reviewed, at what developmental cycle stages the reviews will be conducted, procedures for arranging reviews that are not prescheduled in the project plan, and so on. There should also be documented procedures for selecting a chairperson, or moderator, for the review, and all other activities that must be carried out prior to the review, such as establishing guidelines, collecting and distributing documents for review, establishing checklists, and so on. Reviews may involve a variety of employees: QA engineers, developers, technical experts, and in some cases, a representative for the customer. Review participants may serve in a number of roles in a design review. Some examples of roles in a code or design review might be:

- *Moderator*—The moderator manages the review, insures that it proceeds smoothly, takes notes, and completes forms during the review.

- *Author*—The author of the reviewed design document or source code should be present. Generally, the author will take notes during an informal review.

- *Reviewers*—Reviewers are other members of the development team or organization who offer suggestions and inspect for errors.

It is important to inform review participants of any errors found during a review. They must know that these errors will not, in any way, impact the evaluation of an employee's personal performance in the organization. Reviews should be designed to uncover design and coding errors, not to measure the skills of a specific developer.

The supplier should develop standard forms to be completed during each formal review (see examples in Appendix D). Minutes of all review sessions must be kept, and any action items should be listed. It is also a good idea to hold a follow-up meeting to insure that all action items have been appropriately resolved. In some cases, management leverage might be required to complete action items.

5.7 TESTING AND VALIDATION

Testing and validation suggests that the supplier evaluate a system or system component at the end of the appropriate development phase to insure that it meets all specified requirements. Testing can occur at a variety of stages during the software development life cycle:

- Unit level—involves testing one unit or module of the program, usually comprised of anywhere from 50 to 500 lines of code. At this level, it is usually difficult to use the requirements specification to generate test cases.

- Integration level—involves testing the interaction of program units. Test cases must be generated based on the program architecture and structure.

- System level—involves testing the complete system.

- Acceptance level—involves testing the delivered software product to the customer's requirements specification. This level is often conducted on site with the customer, and is sometimes called field testing.

The supplier must generate a document describing the test plan for each software project. According to the 9000-3, this plan may be an independent document, part of another document, or it may be comprised of several documents.

Test Planning

The supplier must review and approve all test plans and procedures before beginning the testing activities. There should be test plans for each level of testing conducted in the organization. The types of tests performed at each level should also be documented. For example, at a system level, in addition to functional and performance tests to evaluate the ability and stability of a system, most companies also conduct tests to evaluate the

user interface and ease of use of a system. In addition, when testing iterative releases of a system, regression tests should be implemented to insure
that previous functionality was not broken by new functionality. All attributes of the test environment must be outlined in the documentation. For
example, the documentation should describe all hardware or additional software (such as automated testing tools) to be used in testing. The documentation should also define the criteria on which the completion of the test will
be decided. Plans for testing the user documentation should also be included
in the test plan. The responsibilities and personnel involved in testing should
be documented. The test plans must be reviewed and approved before any
testing activities begin. Records of this review and approval must be added
to the quality documentation.

Testing

The supplier must document all testing processes as test cases. The
test cases used during testing must be documented and maintained as records.
All test results should be recorded. Any problems discovered during testing
are recorded and reported to the appropriate personnel. All problems should
be tracked until they are resolved. Once a problem is corrected, the module
and any modules affected by the modification must be retested. The software
and hardware configurations of all testing equipment (hardware, operating
systems, and so on) must also be referenced in testing records.

Validation

It is the responsibility of the supplier to test the complete product
before delivery to the customer. If possible, the supplier should test the
system according to the application environment as outlined in the contract. The supplier must have documented procedures for implementing
validation of the complete system. Records of all test results and any problems detected during validation should also be maintained.

Field Testing

In some cases, the contract may require that field testing be implemented. If this is the case, the supplier should address the system features
tested in the field environment. The supplier should also note the responsibilities of both himself and the purchaser in executing the field test. The
customer's environment should be returned to normal following any field
tests. Some examples are:

- return hardware configurations to their before-test status
- restore network configurations to their before-test status
- remove all software installed for the test
- verify that the customer's system is working normally

The supplier must document procedures for executing a field test, keep records of field test results, and record any problems detected during the test.

5.8 ACCEPTANCE

As mentioned previously, the contract should include some criteria to be used in judging acceptance of the completed system. When the supplier has validated the product and is ready to deliver it to the customer, some type of acceptance test (as defined in the contract) must be performed to demonstrate the conformance of the system to specified requirements. The acceptance test may be carried out by the customer alone, or it may be implemented by the purchaser and customer working as a team. The procedures for handling any problems detected during this phase are documented by the supplier. Records will document the results of testing and any problems identified. If software is not developed for a specific customer, an internal acceptance test conducted by independent personnel should determine whether or not the product is actually ready for release. A separate testing organization will often perform internal acceptance testing.

Acceptance Test Planning

The acceptance test plan should include a schedule for all testing activities, the procedures used for evaluating the software (test cases), the specified hardware and software resources used during testing, and the final acceptance criteria (defined in terms of function and performance requirements). The plan should be generated cooperatively by the supplier and the customer. The test plan should be reviewed and approved to insure that it will exhaustively test all aspects of the system and that all specified resources will be available for testing. The supplier should maintain records of test plan review and approval. It is the purchaser's final decision whether or not to accept the software product (within the limits established by the acceptance criteria as stated in the contract).

5.9 REPLICATION, DELIVERY, AND INSTALLATION

This element maps directly to the handling, storage, and packaging requirements of ISO 9001. The supplier must meet these requirements, as

stated in 9001, in addition to addressing the following special considerations (as outlined in 9000-3) for software.

Replication

All replication activities should be completed prior to the delivery of
the final product to the customer. The supplier should document the replication process, taking the following issues into consideration: the number
of copies of each software item to be delivered, the type of media for each
software item (diskettes, tapes, CD ROM, and so on), including a label
stating the format and version number. The supplier must also consider
the replication of all associated manuals and user's guides. All copyright
and licensing terms should be addressed and agreed to by both the supplier and the customer. The supplier should also describe its policy for
storing master and back-up copies of the software, including plans for
disaster recovery. Finally, the supplier should state its obligation in terms
of the time period for supplying copies of the software to the customer.
The supplier should reference the contract for all replication-related
obligations.

Delivery

The supplier will have documented procedures for insuring that all
delivered software products are complete and correct. The procedures should
insure that the current version is shipped. The supplier should also maintain records of all deliveries, including release notes, references to the
purchaser's software and hardware configurations, and a master index.
Any problems discovered after delivery are to be noted.

Installation

The supplier will insure that all responsibilities, whether of the supplier or the purchaser, are clearly documented in regard to installation.
The documentation should consider such things as an installation schedule, access to the customer's facility (security badges, passwords, and so
on), resources (personnel, equipment), and a procedure for formally approving each completed installation. ISO 9000-3 assumes that the supplier
will go to the purchaser's facilities and complete the installation. The supplier should have some means for validating the system once it is installed
on the purchaser's equipment.

Customer training might be another issue linked to delivery. All training requirements should be clearly specified in the contract.

5.10 MAINTENANCE

The maintenance phase of the product life cycle is much more significant in a software development environment than in a manufacturing environment. It is not unusual for the customer to request changes and enhancements to the completed product. In addition, bugs in the product are almost always discovered and must be corrected. Changing hardware and software requirements generally require that software products be upgraded. The supplier will clearly outline all maintenance responsibilities in the contract. The contract should address such issues as the length of service provided and what parts (programs, users guides, specifications, and so forth) of the software will be maintained. According to the standard, maintenance activities for software are typically classified into the following categories:

- *Problem resolution*—"The detection, analysis and correction of software nonconformities causing operational problems; temporary fixes may be used to minimize downtime and permanent modifications carried out later."

- *Interface modification*—"May be required when additions or changes are made to the hardware system, or components, controlled by the software."

- *Functional expansion or performance improvement*—"Functional expansion or performance improvements may be required by the purchaser in the maintenance stage."

Software nonconformities detected in the maintenance phase should be investigated using root cause analysis. This will provide opportunities for both corrective and preventive actions.

The supplier must document all maintenance procedures, including the type of maintenance activities to be performed, which aspects of the product will be covered by those activities, and who will be responsible for performing the maintenance. The supplier should also document how modifications and enhancements will be incorporated into the software product. Maintenance ties in closely with the concepts of quality in servicing which are covered in the next Chapter.

Maintenance Plan

In addition to the maintenance requirements outlined in the contract, the supplier should prepare a maintenance plan for each software project.

This plan should be subject to approval by both the supplier and the purchaser before development begins. The plan should cover the following topics:

- *Scope of maintenance*—The parts of the program and associated documentation that will be maintained.

- *Initial status*—A definition of the initial status of the product approved by supplier and purchaser.

- *Support*—A description of how the software support will be organized—for example, a service representative for the supplier and purchaser, and a method for handling unexpected problems (such as a customer telephone support line). If necessary, the supplier should identify personnel and resources to be used for maintenance activities.

- *Maintenance activities*—A description of the types of maintenance that will be provided—for example, upgrades, problem resolution, enhancements, and so forth. All maintenance activities should be carried out in accordance with the specified procedures for software development. For example, if a customer requests a functional enhancement to the software, a description of the requirement should be added to the requirements specification, a design document should be created and reviewed, the design should be implemented according to standards and methodologies, a low-level code inspection should be implemented, and the enhancement should be tested—at the unit, integration, and system levels.

- *Records and reports*—Records of all maintenance activities will be kept and stored. The supplier should develop a template for maintenance reports (see sample in Appendix D). The procedures for submitting these reports are to be agreed on by both purchaser and supplier and should be documented in a procedures document.

According to ISO 9000-3, maintenance records should include the following: lists of requests, the current status of problem reports, the employee or outsider responsible for correcting problems and implementing modifications, priorities, results of the corrective actions, and statistical data. These records can then be used for evaluating the efficiency of the software product, and the overall quality system.

Release Procedures

It is very important that the supplier and purchaser agree in advance on procedures for incorporating changes and enhancements into the baseline software product. Release coordination can become difficult and confusing when a product is constantly being modified and enhanced for the customer. A predetermined procedure for making product releases and controlling revisions is vital. These procedures should address the decision to send a "patch," versus sending an upgrade or newer version of the complete software program. The procedures should also indicate whether or not new user documentation will be necessary for each enhancement. The documentation should also define different types of releases or revisions, according to their frequency and impact. The supplier must document his or her methods for notifying the purchaser about changes to the software and for confirming the changes. The supplier must also require records describing all changes to the software, including service locations, customers, products, and so on.

Older versions of software and user documentation can become a headache to both the supplier and to the customer. The supplier, in cooperation with the customer, must develop a plan to eliminate older versions. The supplier may decide to have a separate release policy document, or it may be included in the maintenance plan.

QUALITY SYSTEM SUPPORTING ACTIVITIES

The following sections describe quality assurance activities taking place throughout the development life cycle. They are not specific to any one phase of the product life cycle.

6.1 CONFIGURATION MANAGEMENT

Because of the volatile nature of the software development process, it is necessary to maintain some type of control over the entire product life cycle. Everything from requirements specifications to design documents to the finished software product are inevitably altered during the life of the project. Configuration management is a system for tracking and managing software development. Through the use of a global database, configuration management allows developers to track various releases of a software program. In addition, developers can generate change histories, as well as identify the official versions, for a single module of code. Configuration management also provides some level of administrative control. It allows project managers to keep track of the status of a project, the amount of time developers are spending on each software module, product changes, and so on. The system also controls the development environment to prevent two developers from simultaneously updating the same piece of code and it notifies developers of changes in the specifications, design, or program.

Documents which are critical to the development life cycle, such as requirements specifications and design documents, should also be placed under configuration management. This prevents changes from "slipping through the cracks" and insures that everyone is aware of the current status of the project at all times.

According to ISO 9000-3, the configuration management system should be able to "identify each version of each software module, identify versions of the complete product, identify the status of ongoing and completed software development projects, control simultaneous updating of a software module by more than one person, provide coordination for updating multiple products in various locations, track all changes—from the initial change request to the implementation of that change."

Typically, a software company has a change request review board that meets frequently to review new change requests and examine the status of pending change requests.

Configuration Management Plan

The supplier must document all plans for configuration management. A plan should describe how the organization will use its configuration management system, as well as each employee's individual responsibility with regards to the CM system. In most development environments, a configuration manager, whose sole job is to oversee the configuration management system, is designated. The supplier should also outline what types of configuration management activities will be carried out in the organization. The plan should also specify what types of configuration management tools, methodologies and techniques will be used. There is a wide variety of configuration management packages on the market. The supplier must determine what level of management is necessary for the organization. The plan should also document which items will be placed under configuration management and at what stage in the development life cycle each item will be controlled.

Configuration Management Activities

The supplier must document all procedures for identifying and tracking versions of a single software unit. The procedures should address the identification of items from the requirements specification stage up to and including the delivery stage. At any stage in the development life cycle, the configuration management plan should allow the developer to trace all specifications, interfaces, development tools, and associated documentation that relate to the specified software item.

The supplier must clearly document all procedures for managing change to software items. This documentation should include procedures for approving any suggested changes (which includes assessing their necessity and impact on the remaining product), documenting changes, and authorizing the implementation of these changes. It is also very important that the supplier outline procedures for notifying others in the organization about changes and associated effects. For example, the documentation department needs to know when the user interface has changed so they can update the user's guide.

The configuration management plan should also include documented procedures for preparing configuration status reports. If there is a specific configuration manger, this will probably be his or her responsibility. The reports should record the status of all software items, the current list of change requests, and information on all recently implemented changes to the software.

6.2 DOCUMENT CONTROL

Control of documentation is essential to the success of any quality system. All documents relating to the quality of the software product, and to ISO 9000-3, should be controlled. The supplier should document its procedures for documentation control, including listing the documents that will be subject to the control procedures, procedures for approving and issuing procedures, and the change procedures for documents. For more information on implementing a document control system, see the ISO 9001 document, and Chapter 5 of this book.

Types of Documents

The following documents should be subject to documentation control:

- procedural documents which describe the quality system applied to the software life-cycle
- documents defining activities that affect software quality
- planning documents which describe all activities between customer and supplier
- documents which describe the software product, including development inputs and outputs, verification and validation plans and recorded results, user documentation, and maintenance documentation

Document Approval and Issue

The supplier should implement a system which requires that all quality documentation will be reviewed and approved by the appropriate persons. Once approved, documents should be made readily available to those who need them. For example, coding standards and procedures, design procedures, and review procedures should be issued to each software developer, or stored in an area which allows easy access to the documents. In order to prevent any confusion, all out-dated documents must be removed to preclude inadvertent use. The supplier should document its procedures for the approval, distribution, and removal of quality documents. The supplier should also maintain records of the current versions of all documents, distribution lists, and releases. This can usually be achieved through the production and maintenance of a master document list.

Document Changes

This section of ISO 9000-3 is taken directly from ISO 9001. Refer to Chapter 5 of this book for more information.

6.3 QUALITY RECORDS

In order to maintain an effective quality system, the supplier must keep records of all quality related activities. Quality records allow the supplier to monitor the effectiveness of the system, and provide tangible evidence that there is a quality system in place in the organization. This section of ISO 9000-3 is taken directly from ISO 9001. Refer to Chapter 5 of this book for more information.

6.4 MEASUREMENT

The supplier should use some type of metrics to evaluate and control the software development process, and measure the quality of the finished software product for each software development project. There are two main subdivisions of metrics in software development: product metrics and process metrics. Product metrics are used to evaluate the quality of the delivered software product, while process metrics are used to estimate the effectiveness of the development and delivery process.

Product Measurement

While there are no universally accepted measures for software quality, there are a number of metrics that can be used. Some measures of quality may include: the percentage of product failures reported, the percentage of error-free lines of code produced, the execution speed of the program, functionality, ease of maintenance, the degree of portability to various hardware platforms, and so forth. Some of these measurements are easy to collect, while others, such as estimating functionality, may be more difficult. Note that the type of metrics used may vary according to the nature of the product that is being developed. At an absolute minimum, the supplier should keep measurements of product failures according to customer reports and customer complaints. These become inputs to the corrective and preventive action program.

It is not adequate to wait for customer complaints. Feedback should be elicited so that improvements can be made. Dissatisfied customers might not complain, but will take their business elsewhere. Customer surveys are strongly suggested.

No matter what type of product quality measurements are gathered, the supplier should use the results to improve its software products and processes. ISO 9000-3 suggests that the supplier:

- collect and report metrics on a regular basis
- identify the current level of performance on each measurement taken
- take remedial action if metric levels get worse or exceed targets
- establish goals in terms of metrics

Process Measurement

The supplier should also use metrics to evaluate its software development and delivery process. There are a variety of process measurements that can be used. The supplier should use measures relevant to the development processes being evaluated. Some examples of process measurements are:

- recording the time it takes to respond to a customer complaint
- recording the quantity and severity of errors uncovered during a design review
- measuring project progress in terms of meeting deadlines
- measuring the number and severity of problems detected by the testing department

Process measurements aid the supplier in determining whether or not the development process is successful in reducing product nonconformities, and if development is being carried out according to schedules and predetermined quality objectives.

The supplier should describe all product and process measurements, including procedures for performing the measurements, in its quality documentation. Records are kept indicating that these methods have been approved. The supplier's quality records should also include reports of all product and process measurements taken, any problems or deficiencies revealed, and the planned corrective actions.

6.5 RULES, PRACTICES, AND CONVENTIONS

In order for any type of structured system to be successful and remain intact, it must have specific rules, practices, and conventions. A rule is an established, mandatory requirement that must be followed at all times, a practice is a suggested approach to a problem, and a convention is a regularly followed custom. For example, a rule may require that all documents must have the proper approval and signatures before they are placed in the quality system documentation. A practice may suggest holding a high level preliminary design review before proceeding with any coding efforts. A convention may require that all buttons in a user interface be labeled with verbs.

The supplier must document rules, practices, and conventions which will insure that the software quality system is effective. These documented rules, practices, and conventions should apply to the entire quality system. The supplier must be able to measure and verify adherence to each documented rule, practice, and convention, and should keep records of all such measurements. Rules, practices, and conventions should be reviewed and revised when appropriate.

6.6 TOOLS AND TECHNIQUES

The supplier should use tools and techniques whenever possible to make the quality system more effective. Some tools and techniques are used specifically for management of the quality system, such as a computer system which stores, and controls access to, all quality documentation and records. Other tools and techniques are used to improve the actual development process, such as debugging systems, compilers, automated testing tools, and configuration management systems. The supplier should document each tool and technique used, as well as procedures for their selection, review, and evaluation and all replacements and improvements. Tools and techniques should be improved or replaced when necessary.

6.7 PURCHASING

In a software development environment, it is generally necessary for the supplier to purchase additional software, hardware, or services from a third-party vendor. Third-party software or hardware may be included in the supplier's final software product, or it may be used to aid in the development of the supplier's product. Many software applications include off-the-shelf products, such as databases and communications software. For example, many software companies develop database-driven applications, which access information from a third-party database system. While the software company does not develop the actual database software, it is often provided with the product. In some cases, the supplier may purchase sub-contracting services from a third-party organization, or bring consultants into its own organization. In any case, all purchased products or services must conform to the supplier's specified requirements. Depending on the supplier's specifications, third-party companies may need to meet certain professional criteria, agree to an audit, complete a questionnaire, or have a documented and superior quality record.

The supplier should document its procedures for purchasing third-party products or services. These documents should also describe all products or services. Records of all such purchases should be kept and reviewed to insure that all specified requirements are being met. For example, if the supplier will only purchase products or services from organizations registered under ISO 9000, the records should indicate that all third-party companies are registered.

Assessment of Subcontractors

The supplier must also have specified requirements for the selection, use, and evaluation of subcontracting personnel. This element is taken directly from ISO 9001. Refer to Chapter 5 of this book for more information regarding the assessment of subcontractors.

Validation of Purchased Product

If the supplier chooses to subcontract development work, this work should be subject to validation by the supplier. Often, the subcontractor may be required to participate in design reviews and code walk-throughs, in accordance with the supplier's quality system. Acceptance testing of the subcontractor's product should also be required by the supplier. All requirements should be clearly outlined in the subcontract. Whether or not the subcontractor verifies its own product, or even if the customer insists on verifying the subcontracted products, it is still the responsibility of the supplier to verify all subcontracted work.

The supplier must have documented procedures for evaluating and validating all subcontracted service and products. The supplier keeps records of any reviews or related development activities, acceptance tests, and problems found in the subcontractor's product.

Included Software Product

It is common for a customer to require that the supplier include some type of product (such as code or data) in the completed software package. If the supplier is required to include or use a third-party product, or a product supplied by the customer, there should be documented procedures for validation, storage, protection, and maintenance of the product. The supplier must also consider how it will support such products in regard to maintenance of the delivered software. The supplier must insure that the customer's responsibility is documented in the contract. Issues such as what type of software will be provided, schedules for providing the software, responsibilities of the supplier and purchaser, and testing of the included product should all be addressed in the contract. If a software company is developing a program that will interface with the customer's CIS (customer information system), the responsibilities of the customer should be stated in the contract, for example, the date that the source code and program architecture will be provided to the supplier.

The supplier should include a description of all included software products in its quality documentation. There should also be documented procedures for testing, storing, and maintaining the product and quality records that reflect validation activities, proper storage, maintenance activities, and any problems with the included software product. Customer supplied product has been previously discussed in reference to element 4.7 from ISO 9001 (see Chapter 5).

6.9 TRAINING

In order for any organization to perform well and produce a quality product, the employees must be adequately trained. The supplier should identify training needs in its organization. Training need assessments may be carried out on an annual basis. As organizations progress, training needs change. For example, the acquisition of new tools and hardware produces new training needs. Any individual in the organization whose job can affect the quality of the delivered software product must be adequately trained. The supplier must be able to show that employees have the appropriate education, training, and/or on-the-job experience required to perform well in each position.

The supplier should have documented procedures for identifying training needs. This may involve an annual process that begins with updating resumes and reviewing these resumes for deficiencies. The supplier should have documented procedures for filling training requirements. Whether training is provided in-house or out, records must be kept.

Chapter Review Questions

1. Which two life cycle stages are most important in a software development environment?
2. How was software quality typically addressed in the past?
3. What is the purpose of ISO 9000-3?
4. Identify two registration options for software development companies.
5. What is TickIT?
6. What does SQSR stand for?
7. What is the purchaser's responsibility in software development?
8. List some aspects of software development that should be addressed in the quality manual.
9. What issues should be addressed in the quality plan for a project?
10. Identify the five basic phases of a software product's development life cycle.
11. List some examples of quality-related issues that should be addressed in the contract.
12. List two reasons why requirements specifications are so important.
13. What is the purpose of the development plan?
14. Why is the design phase of the software development life cycle a critical quality issue?
15. What are the two basic concerns that must be addressed in written implementation procedures?
16. What are the two types of reviews that must be held during the development life cycle?
17. What are the four stages of development where testing must occur?
18. What are some issues that must be addressed in field test procedures?

19. What are some issues that must be addressed in replication procedures?
20. What types of things must be documented in quality records for software deliveries?
21. What are the three basic categories for software maintenance activities?
22. What are some maintenance issues that must be addressed in the contract?
23. List the five areas that must be covered in a software maintenance plan.
24. What types of concerns must be covered in the release procedures document?
25. What are some of the purposes of configuration management?
26. What should be documented in the configuration management plan?
27. List the types of documents that must be subject to document control procedures.
28. List and define the two types of metrics used to measure software development quality.
29. Define the terms rule, practice, and convention.
30. List some ways in which a third-party vendor or contractor can be verified.

8

ISO 9004-2: QUALITY IN SERVICE

INTRODUCTION

The service sector is growing and so are customer expectations of that sector. Quality service can be every bit as important as quality goods. In fact, the quality of services offered by manufacturers can overshadow the quality of their goods. This is especially true when the quality of services lags behind customer expectations. This chapter is important for companies and organizations in the service sector, but it is also relevant for those manufacturers that provide their customers with both product and service. As shown in Figure 8-1, most organizations offer a combination of product and services. The quality of both must be acceptable for the organization to be viewed favorably by customers.

This chapter is based on *ISO 9004-2: Guidelines for services* which builds on the quality management principles delivered in the ISO 9000 to ISO 9004 series. Applying these principles to service provides:

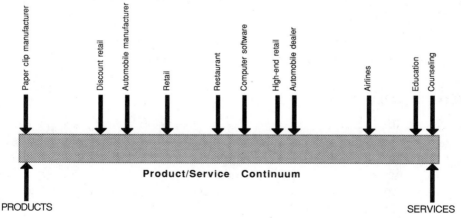

Figure 8.1 The product/service continuum

- service improvements and increased customer satisfaction

- better productivity, efficiency and cost reduction

- improved market share

These benefits can be achieved by attending to the human aspects involved in providing services:

- managing social processes

- regarding human interactions as a crucial part of service quality

- recognizing the importance of the customer's perceptions

- developing the skills and capabilities of personnel

- motivating personnel

In the service industry:

1. One instance of poor service can negate many instances of acceptable service or even exemplary service.

2. It costs an organization significantly more to correct an instance of poor service than it does to initially deliver proper service.

3. It costs an organization much more to establish a new customer than retain an old one.

4. Customer assessment of any nonconformity is often immediate.

5. The best place to apply quality control is to the process that delivers the service.

6. Most instances of poor service are due to inadequate procedures, instructions, resources, environment, or personnel training.

SERVICE QUALITY DESIGN

The critical factor in designing a service is to concentrate on the needs of the customer with special attention to what the customer feels is most important. This seems to be sensible but is often forgotten, especially in those organizations where the focus is on product rather than service. In some instances, management views the service operation as a necessary evil that is tolerated in order to be able to sell the product. This is an unfortunate point of view because the customer might be more concerned with the service than with the product. Consider the example of two restaurants: both offer equally fine food but one has poor service. Dining out is a social event for most people. When the service is bad, the experience is bad. If the two restaurants are competitors, there is little doubt as to which one will fare the best. In fact, most people would rate the food better at the restaurant that provides good service, because customers tend to view product and service as a whole.

Market Research

The marketing division of the service organization will determine the need of, and promote the demand for, a service. Surveys and interviews can be used to determine:

- customer needs and expectations
- quality of service and reliability expected
- availability and competitor activities
- complementary services
- relevant legislation (i.e., health, safety, and environmental)
- relevant standards and codes
- safety and liability issues
- relevant environmental concerns
- project profitability
- market changes and new technology

New technology is changing organizations in the service sector. Consider the use of computers to help customers visualize the potential of a service (examples include hair styling, remodeling, and cosmetic surgery). In fact, new technology is creating innovative areas for service and unique business opportunities.

Service Brief

The results of market research and analysis, and the supplier's obligations should be incorporated into a *service brief*. This brief defines the customers' needs and the supplier's capabilities as a set of requirements and instructions that form the basis for the design of a service. The service brief should also reference warranties, quality documentation, and any regulatory requirements. After the service brief has been established, the design process will produce three categories of specifications:

1. the service itself

2. the means and methods used to deliver the service

3. the procedures for controlling the quality of the service and its delivery

Figure 8-2 shows how the design process fits into the service quality loop.

Organizations should pay particular attention to the design of the service delivery system. Some services can be segregated into customer contact and customer non-contact processes. Often, the service delivery can be made more efficient and effective by uncoupling these processes and by using different designs for each. For example, the customer contact process will be better handled by personnel with discerning human skills and the appropriate training. The non-contact process can be handled by personnel

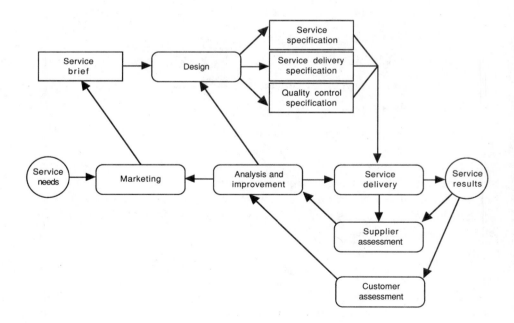

Figure 8.2 Service quality loop (adapted from ISO 9004-2)

with the requisite technical skills. In some cases, a non-contact process might be a candidate for automation.

The service specification should contain a complete and precise statement of the service to be provided, including:

- a description of those characteristics subject to customer evaluation
- a standard of acceptability for each characteristic

The service delivery specification should describe the methods to be used in the delivery process, including:

- a description of the delivery characteristics that directly affect performance
- a standard of acceptability for each delivery characteristic
- the physical and material resources required
- the number of personnel required
- the skills and knowledge needed
- the reliance on subcontractors

The service specifications, delivery specifications, and quality specifications are interdependent and should be fully interactive during the design process. One suggested method to track these interdependencies and interactions is flow charts. Management should assign design responsibilities for:

- planning, preparation, validation, maintenance and control of the specifications
- identifying the facilities, resources, products and services that are needed to support the specifications
- implementing design reviews for each phase of the design
- validating that service delivery meets the service brief requirements
- revising the specifications in response to internal and external feedback

A design review should be conducted and documented at the end of each design phase. The design should be checked for consistency with the service specification, the service delivery specification, and the quality control specification. The design review should be attended by all departments whose functions affect the quality of the service phase being reviewed. This is the time to verify that all customer requirements have been addressed and that the quality control specification is adequate. When possible, problem areas should be anticipated, and contingency plans and/or design changes should be considered for these areas.

Design Changes

The design specifications should not be changed without due cause and careful consideration. Design change control procedures should be followed to insure that:

- the need for change is identified and verified
- the change is analyzed for its impact on the service offered
- changes to the specifications are planned, documented, approved, implemented, and recorded
- representatives of all affected functions participate in the redesign and approval process (this could include customers)
- the impact of all changes are evaluated to insure that service quality is not degraded
- all personnel receive needed information and instructions
- customers are informed when changes will affect performance characteristics

Service Changes

New and modified services should be validated to insure that they are fully developed. The validation process should be defined, planned, completed, and documented prior to offering the service to customers. The validation process should determine that the:

- service is consistent with customer needs
- delivery process is complete
- required resources are available, particularly materials and personnel

- applicable codes, standards, practices, drawings, and specifications are satisfied

- all personnel receive needed information and instructions

- information needed by customers is available

Services should be periodically revalidated to insure adherence to the design specifications and to identify potential areas for improvement. Revalidation should be a planned and documented activity based on actual field experience. The major source of information will usually be the customers and the personnel providing the service. It will be necessary to determine the impacts caused by any changes in procedures, personnel, customer needs, market conditions, competition, or available technology.

MANAGEMENT FUNCTIONS

Management should develop and commit to a quality policy that addresses the:

- specific services to be provided

- image and reputation of the company

- objectives for service quality

- approach to be used in pursuit of the quality objectives

- role and responsibilities of those developing and implementing the quality policy

The quality policy must be translated into specific goals and actions. Some examples of quality objectives and activities are:

- to develop clear definitions of customer needs

- insure that the service design and delivery process address customer needs

- employ controls and preventive measures to avoid customer dissatisfaction

- to decrease costs and increase performance

- create a company wide commitment to quality service

- use continuous review to maintain and upgrade standards

- prevent adverse effects on customers, society, and the environment

Management must also insure that the quality policy is promulgated, understood, implemented, and maintained.

It is the responsibility of management to assign responsibilities and make planning considerations in the areas of:

- the introduction of the service and its eventual withdrawal

- relevant safety issues

- relevant liability issues

- relevant environmental issues

- potential failure points and the consequences of failures

- contingency plans

- variations in demand for the service

Any of the above issues can be critical for an organization. As an example, consider variations in the demand for a service. Response time is very important to customers. In some cases, it is the most important criterion used by customers to gauge performance. In our society, our impatience has become one of our distinguishing characteristics. Every other aspect of a service can be above average, but if customers have to wait too long, they might look to another organization. A service provider must invest in sufficient resources to be in a position to exceed or meet customer demands but must not over invest. Some service providers have successfully used their customers to help meet demands (i.e., self-serve salad bars and gasoline stations).

A service provider can use capacity requirements planning to determine if it is possible to meet customer demands. Some organizations use aggregate planning. In this method, activities that share resources are totaled (aggregated) to determine needs. Then the organization determines its aggregate resources. An example of an aggregate resource is an organization's available production hours, as determined by the formula:

Monthly Hs. = no. of employees x no. of work days x 8 Hs.—vacation Hs.

After aggregate needs are balanced against aggregate resources, various adjustments can be made to improve efficiency and response time. Aggregate planning can be a waste of time in a small service organization, but in a complex organization, with several divisions that overlap, aggregate planning might yield worthwhile benefits. For example, it might be determined that several floating employees could improve response time at an acceptable cost to service ratio. Another possible output from aggregate planning is new and novel ways to locate and share physical resources among service sites.

Management might also investigate various customer scheduling tactics such as:

- Appointment systems. Examples include doctor's offices and auto-service shops. This approach can be ideal in non-emergency servicing and slack time can be built in to accommodate emergencies and unforeseen difficulties.

- Reservation systems. Examples include hotels and restaurants.

- Strategic pricing. This approach is used by telephone and power companies. Such organizations experience decreasing profits when meeting peak demands because they must invest in additional equipment which then sits idle during off-peak periods. Strategic pricing encourages a shift of customers to off-peak periods.

Service scheduling should make allowances for every part of the process. For example, repair services should include time for testing after the repair to verify that the repairs are complete and correct. Service organizations can sometimes gain efficiency and better meet customer demands by changing employee schedules and assignments:

- Flextime, where the employees choose start and stop times to achieve a 40 hour week.

- Flextour, where the employees choose start and stop times but are restricted to five, 8-hour days.

- Compression, with four, 10-hour workdays.

- Staggering, where employees choose times from an approved list.

- Split shifts, where employees work two shifts per day.

- Duty tours, where employees are on-call for a period (perhaps 24 hours) followed by an off period.

- Part-time and permanent part-time employees.

- Floating workers, who can be moved between or among tasks and/ or locations according to demand.

- Mixing, which uses various combinations of the above.

Service responsiveness can sometimes be improved by decentralizing purchasing. When service managers are authorized to purchase materials or services, they are often in a better position to provide what their customers need in a timely fashion. However, all purchasing *must be carefully controlled* because of its potential to impact the quality, cost, and efficiency of the services offered. Subcontractors should become partners as discussed in Chapters 2 and 5. The minimum requirements for the control of purchasing are:

- documentation in the form of purchase orders and additional records if needed to provide traceability

- qualification of subcontractors

- use of nonconformity records

- agreement on quality requirements with all subcontractors

- agreement on methods of quality assurance and verification methods

- provisions for resolving quality disputes

- incoming product and service controls

- records of the quality of incoming product and service

To properly assess subcontractors, an organization should consider:

- on site inspections and evaluations of the subcontractor's quality system and capabilities

- its history of dealing with the subcontractor

- evaluation of samples

- comparisons with other subcontractors

- the experiences of others

Management Review

There should be provision for both formal periodic and independent reviews of the quality system to provide for its stability and continuing effectiveness. Particular emphasis should be placed on the need and opportunities for improvement. These reviews should be documented and conducted by management . Relevant sources of information include:

- findings of service performance analysis in the areas of effectiveness, efficiency, and customer satisfaction

- internal audits conducted by personnel independent of the areas being audited, to determine how well stated objectives are being met by established procedures, with particular notice to the service delivery specification and the quality control specification

- changes brought about by new ideas, new technologies, shifting market strategies, and changes in the business environment

QUALITY ASSESSMENT

The quality of service is determined by both tangible and intangible factors. One way to view the quality factors is to divide them into technical factors and human factors. The technical factors are objective and they can be measured. These factors include:

- Competence—After servicing, the car horn worked. This category can be called "accuracy" in other service sectors—the shoes you ordered were delivered in the correct size and style but the color was wrong.

- Reliability—The horn failed again, one week after it was serviced.

- Deliverability—The service meets the provider's advertising claims and all provisions in the service contract.

- Timeliness—The shop scheduled the car the same week the customer called and then met that schedule.

- Accessibility—The service is available when and where the customer can use it.

- Cost—The charge was equal to the estimate which was competitive with estimates from other service providers.

- Workmanship—The horn works but there is a scratch on the fender.

The intangible factors are subjective and they are almost totally dependent upon the interaction of the customer with the service personnel. They are human factors, and they include:

- Courtesy—The personnel are friendly and recognize regular customers.

- Credibility—The personnel appear to be competent. They seem to have the necessary tools and materials to furnish the service.

- Communication—The service is explained in adequate detail and in terms the customer can understand. The customer's questions are answered in a satisfactory fashion.

- Appearance—The personnel, the equipment, and the service facility are neat and clean.

- Attitude—The personnel do not "bad mouth" a product, their company, other providers, or the manufacturer. Their behavior is positive and builds the customer's confidence.

- Honesty—The personnel do not exaggerate, make excuses, or make outlandish promises. Estimates are not "deflated" to make a sale.

- Empathy—The personnel seem to be considerate of the customer's problems and special needs.

It's a good idea to apply appropriate control measures to the technical factors. Since they lend themselves to objective measurement, quality control specifications can be established to insure that the service consistently meets the service specification and the customers' expectations. The quality control design involves:

- identification of the key activities in each process which have a significant influence on the specified process

- analysis of the key activities to select the characteristics whose measurement and control will insure service quality

- defining the evaluation methods

- establishing measures to control the characteristics within specified limits

The human factors are determined by the personnel at the customer interface. Selection, training, motivation and the judicious use of customer feedback are the primary management tools to control the quality of these factors:

- select personnel on the basis of capability to meet defined job specifications

- create a work environment that fosters security and excellence

- provide opportunities for employee creativity and greater involvement

- provide training and retraining

- insure that employees understand and appreciate how their function impacts quality

- make the customer the focus

- give due recognition for employee contributions and suggestions

- assess and improve the motivational factors

- implement career planning and personnel development

Personnel delivering the service will often have an opportunity to obtain customer feedback. Depending on the type of service and the circumstances, it might be possible to assay the customer's satisfaction with the service. With the proper training and skills, employees can gain invaluable information from customers.

Customer satisfaction in the area of human factors can also be determined with follow-up communications. This can take the form of a reply card, questionnaire, or a telephone call. Such efforts should be carefully designed to probe the vital areas in ways that do not annoy customers or waste their time. Customers tend not to volunteer their assessments of service quality to the service organization. If dissatisfied, they often cease to deal with the supplier without giving notice or specific complaints that would enable corrective action to be taken. Therefore, relying on customer complaints as a measure of customer satisfaction can lead to improper conclusions.

When sampling customer reactions, the service supplier should focus on how well the service brief, specifications, and the service delivery process met the customer's needs. The supplier should make comparisons between customer assessments and internal assessments. Discrepancies could point to inadequate specifications, processes or measures in providing the service.

Some service organizations use a system of *exception reporting*: management is informed of all those instances where delivered service does not comply with the stated requirements, objectives, or customer expectations. The exception reports are generated by the service personnel, by the personnel engaged in customer follow-ups, and those receiving customer complaints.

Customer Input

Persons with direct customer contact are an important source of information. Customer communication involves listening to them and keeping them informed. Any communication problems should be remedied as soon as possible. These problems should also be carefully analyzed to determine if there are deficiencies in the service delivery process. The critical areas of customer communication include:

- an accurate description of the service including its scope and the time frame of its delivery

- a clear statement of all prices

- the relationships among service, delivery, and price

- potential problems and how they will be solved

- what is expected of the customer

- the ways the customer affects the quality of the service

- provision for materials and facilities to support communications

- any secondary effects that a service might produce

In order to establish, preserve, and advance the quality of service the supplier must evaluate the delivery process by:

- measuring and verifying service activities to avoid undesirable trends and customer dissatisfaction

- self inspection performed by the service delivery personnel

- sampling customer reactions

DOCUMENTATION AND TRAINING

As it is for quality systems for manufacturing and software, the documentation methodology is a central and critical element for controlling the quality of service. All of the service specifications, requirements and provisions should be defined and plainly stated. The *quality manual* is the premier document and it should contain:

- the organization's quality policy
- the organization's quality objectives
- the structure of the organization
- the assignment of responsibilities
- a description of the quality system
- the quality practices
- references to those documents that control and certify quality

The quality plan describes the specific practices, resources, and the sequence of activities for a particular service. The quality procedures specify the purpose and scope of the activities used to meet customer needs. They also define the method of control and documentation. All quality procedures should be agreed upon, be clearly stated, and be accessible to the appropriate personnel.

Quality records should be accessible and verified as being valid. They should be retained for a designated time and protected from loss and damage. These quality records provide information in areas such as:

- service performance, including customer feedback
- instances of nonconformity and actions taken to satisfy customers
- corrective actions to eliminate reoccurrence of nonconformity
- potential areas for improvement, including matters of efficiency
- analyses for quality trends and the need for improvement
- the quality of performance of subcontractors
- personnel performance, including training needs
- competitive comparisons
- calibration and maintenance records

Quality documents must be controlled. The specifics of control include the issues of:

- identifying documents as being controlled

- document preparation, dating, authorization, release, and distribution

- review and revisions

- the disposal of obsolete items

Training

Human resources are the most important ones for any service organization. Education and training are investments that are necessary for both maintaining and improving quality. Before training systems are designed, an assessment of requirements and employee qualifications should be performed. An effective system includes methods to verify that employees have received and understand the information that they need to perform effectively. Some areas for education and training include:

- safety concerns and regulatory issues

- quality concepts and quality system design for management, including evaluation methods and how to determine costs

- specific technical skills for personnel involved in the service delivery process

- information regarding the general quality policies and procedures of the organization

- customer expectations and how to deal with customers (human skills)

- orientation courses for new personnel and refresher courses for established personnel

- process control, data collection, data analysis, problem identification, problem analysis, corrective action, improvement, team methods, and communications

Team efforts and information sharing are activities that enable an organization to function smoothly on the inside and interface effectively with its customers. Some organizations have unfortunately learned that not paying attention to this area fosters the growth of rumor mills which tend to be disruptive and destructive. Some human beings, when lacking information, invent news, and they tend to invent far more bad news than good news. Herein lies a significant opportunity for management to be visible in a meaningful and positive way. Some of the kinds of team efforts and informational meetings that can work effectively include:

- management briefings
- information exchange and idea sharing
- the handling and prevention of nonconformities
- distribution and discussion of new and revised documents
- new technology
- discussion of common problems and solutions
- planning sessions
- areas for improvement
- news about changes in the market, the competition, etc.

Large organizations often use newspapers, bulletins or flyers to disseminate information among their various divisions and departments.

Chapter Review Questions

1. List some reasons why the quality of service can be critical to the well-being of a company that provides goods.
2. Identify the causal factors of poor service quality.
3. Can the quality of service affect the perceived quality of goods? Cite some examples.
4. Is market research any less important in the service industry than in other industries? Why?
5. What are some of the things to be learned from market research?
6. Describe what goes into the service brief.
7. List the three categories of service specifications.
8. What are the major functions of the design review?
9. Who should participate in the design review?
10. What must be considered before the design of a service is changed?
11. Identify the elements in the service design validation process.
12. Discuss how aggregate planning might be used and some possible results.
13. Discuss several scheduling systems that can improve response time and/or improve efficiency.
14. Is response time a critical measure of service quality? Why?

15. Elaborate on the arrangements with subcontractors that are necessary to maintain quality standards.
16. Identify the sources of information used in a management review.
17. What are the technical factors for quality assessment in the service industry?
18. What are the intangible factors for quality assessment?
19. How can quality control be applied to the technical factors?
20. How can quality control be applied to the intangible factors?
21. Name some ways to measure customer satisfaction.
22. What is exception reporting and how is it used?
23. List the things that a customer should know.
24. What should be in the quality manual of a service provider?
25. Identify some quality records that would be appropriate for service providers.
26. List some important areas for employee training.

9

IMPLEMENTATION AND AUDIT PREPARATION

INTRODUCTION

One of the most valuable outcomes of an external audit is an objective view of the organization. For some companies, it is a form of reality check that is long overdue. We all know that it is very difficult to be truly objective about one's self, one's family, or one's company. Outside viewpoints, which come from informed sources and are properly structured can provide invaluable information and lead to significant improvements.

TOOLS

The best way to implement an ISO 9000 quality system is to start with an independent third-party audit. This audit, called a *benchmark* or *baseline audit,* is a thorough review of the quality system currently in place. The audit compares the current system to the ISO 9000 standards desired, 9001 or 9002, and reports in detail how the quality system fails to conform to these standards. The benchmark audit is frequently a very sobering exercise for the organization, but it does let everyone know the magnitude of the job ahead and helps formulate the project plan by establishing goals.

A good way to represent the project plan of an organization is by using some form of an activity based sequential chart so that its staff can see who is responsible for any given task and when the task is to be completed. It is necessary for the ISO coordinator to insure that all tasks are completed on time. The chart needs to be distributed at high levels of the organization to insure that there is proper exposure to delayed activities.

Baseline Audit

It is important to get an independent third party to conduct the *baseline audit*. In-house personnel can do the audit, but the politics of most organizations can make the results less than candid. Of course, multi-site companies can sometimes obtain an acceptable degree of objectivity using an audit team from a different company location. However, remember that auditing involves special skills and that staff members from out-of-town are not always experts.

The independence of outsiders assures that the audit team will not try to sugarcoat the results. An organization especially needs the cold, hard truth about its quality system if it is ever going to make the changes necessary to succeed. The independent third party should consist of people who not only are familiar with the ISO standard, but have direct knowledge of how the registrars interpret the standard. This experience comes from conducting registration audits. It is also important that the audit team has hands-on experience at implementing an ISO 9000 quality system. There is nothing equal to having been responsible for implementing a quality system when auditing to detect quality system deficiencies.

The audit is a learning experience for the entire organization, so the auditors must have excellent communications skills. The auditors will be talking with those at all levels of the organization, giving verbal briefings to large groups, writing nonconformances, and developing the formal written report. It is helpful, but not essential, for the auditors to be knowledgeable about the processes being audited. It is more important for the auditors to be good at auditing and communicating.

Project Plan

The nonconformances are used as the primary material in a project plan for implementation. They are written on specific ISO 9000 elements for different areas of the operation. These nonconformances make it easy to break down the huge task of creating an ISO 9000 compliant quality system. The ISO coordinator should take the individual noncompliances and sort them into related categories so that discrete tasks can be assigned to members of the ISO implementation project team. These tasks must be given to personnel who have the authority to make the needed changes. Of course, the tasks can be delegated to others, but the responsibility for getting the tasks completed are given to the members of the ISO implementation team.

Corrective Action

The nonconformances are only the start of the quality system implementation process. There should be some thought devoted to how the noncompliances will be corrected. Corrective actions should not be too simplistic in nature, such as addressing only the individual noncompliance. A systemic approach must be used to insure that a band-aid is not being placed on one instance of a widespread problem. For example, if the nonconformance stated that a micrometer was found without a calibration sticker, the corrective action should not only address getting the micrometer calibrated and the sticker attached, but it should also pursue the definite possibility that there are more measurement devices in the work place without calibration stickers. Yet other issues might be addressed:

- Are all measurement devices listed in the calibration data base?
- Are all measurement devices properly numbered?
- Does the calibration system procedure have a method by which all required calibrations are brought to the owner's attention?
- How are new measurement devices brought into the calibration system?
- Are new measurement devices routed through the calibration lab before being put into use?
- Are the calibration labels tenacious enough to stay on the measurement devices for the duration of the calibration period?
- Who has the ultimate responsibility for getting measurement devices calibrated? Is the policy understood throughout the organization?

These more systemic issues must be addressed in all corrective actions to insure that the quality system will comply with the ISO 9000 standard.

Monitoring Progress

It is important to have a way to monitor the progress of the implementation activity. There are a number of graphical techniques that are effective, but the one most commonly used is the Gantt Chart. Its primary function is to show the tasks, the people responsible for the tasks, and the date the tasks are to be completed.

There are many features of Gantt Chart analysis that can enhance the basic chart, but its power is in its ability to show on a simple graph

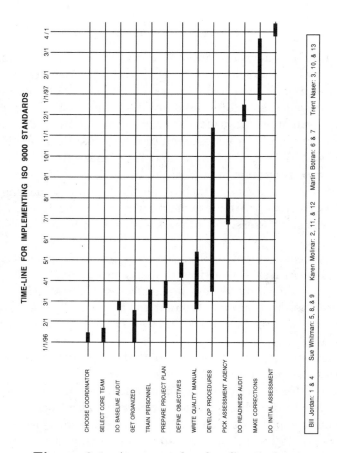

Figure 9-1 An example of a Gantt chart

where the bulk of the activity is required and the progress being made (or not made). There are many good PC-based software programs for project planning that can make the generation of Gantt Charts and other graphs a relatively simple task.

ISO Coordinators

The ISO coordinators in each area should insure that the tasks are delegated to people who are capable of getting the tasks accomplished. They are also the ones who need to follow up on the timing of the tasks. There should be strong communications within this group so that if barriers appear, management can be informed and resources will be put in place to hurdle the barriers. The best form of communication is to hold regular meetings of the ISO steering committee or core team. Since bonuses are often tied to successful ISO implementation, any problems discussed in these meetings will be appropriately addressed. These meetings must be crisply run and facilitated effectively. It must be understood by all that the ISO effort is too important to fail, and to remain efficient and viable in the marketplace, and the health of the company depends upon it.

DISCIPLINE

Once the ISO implementation process starts, it must move inexorably forward. Crises will frequently occur that will make people think that they must delay ISO implementation and address the crisis. This must not be allowed to happen. Momentum is everything in an undertaking the size of ISO. If the focus gets diverted to other activities, ISO will become an afterthought and will not be implemented. Holding management reviews to evaluate the progress of ISO implementation is a very effective way to keep the ball rolling. It also develops the format for management reviews that are a necessary part of the ongoing maintenance and improvement of the quality system (see section 4.1 in Chapter 5).

CLEANUP

All during the implementation process, internal audits should be conducted to insure that the system being put in place complies with the ISO 9000 requirements. Where appropriate, consultants should be used to conduct occasional and rigorous audits. These act as objectivity checks for the internal audits and can legitimize the internal audit findings if area personnel become skeptical about the internal audit process. The noncompliances generated should be used to make mid-course corrections in the implementation project plan. The noncompliances must also follow the ISO requirements for verifying corrective action implementation and effectiveness.

REGISTRATION AUDIT

The registration audit is the culmination of the ISO implementation effort. It must be managed at least as well as the implementation process was managed. There are a number of good firms who conduct registration audits. It is advisable to ask other associates who have been through the registration process, or get in touch with the American Society for Quality Control to investigate which firm might be the best for your business. This should be done at least six months before the actual registration audit is to take place.

A preliminary audit is sometimes recommended by the registration firm to insure that there are no major gaps in the quality system. It would make no sense to schedule a major ISO audit if the quality system is obviously not ready. It just wastes the auditors' time as well as that of those who are being audited, and can also build up frustration and other negative feelings about the whole process. The preliminary audit is a brief overview used by the registration company to review high level documentation and assess the general readiness of the organization for the registration audit. Any major deficiencies are noted and the auditors deliver a brief report that declares what must be done to prepare for the major audit. The preliminary audit should be scheduled two to three months before the registration audit to insure that there is sufficient time to correct any major problems.

AUDIT MANAGEMENT

A well prepared audit plan can make the difference between success and failure of the audit. Insuring that the auditors are well cared-for is not just gamesmanship, but is a common courtesy. If flying in, the auditors should be picked up at the airport and transported to their hotel. If they prefer to rent a car, good directions should be mailed beforehand showing the best way to the company facility and to the hotel. Meeting the auditors for dinner or breakfast is a nice touch.

Once the audit has started, however, the auditors will need time in the evenings to gather their notes, meet for discussions, and write noncompliances. They may not wish to be taken to dinner every evening because the process is very hectic for the auditors and they need some off time to collect their thoughts and regenerate their energy. The hotel chosen should be based on comfort and convenience and not just on price. This is not an area where a company should scrimp. Auditors are human, and they have good days and bad days like the rest of us. If they have a miserable night in a stuffy, noisy hotel, they might not be as charitable when it comes to judging whether a noncompliance is major or minor. The ISO 9000 standard is well-written, but it is subject to interpretation by the auditors. An agitated auditor is not the best person to depend upon for registration.

The audit itself should be orchestrated to minimize any potential problems. It is good to suggest an audit schedule ahead of time, but let the auditors make any modifications. In a certification audit, any area within its scope is subject to being audited, even if it is not on the schedule. Audit trails can lead anywhere, so there is nothing sacred about the schedule if an auditor is following the trail like a bloodhound. Audit escorts, knowledgeable in the areas to be audited, should be assigned for each auditor. Each area to be audited should have a manager ready to guide the auditor and to answer questions about management of the area. Operators in the area will be interviewed, but it is permissible for the manager to help out if the operator is too nervous to answer the questions clearly.

A control center should be set up for all information coming out of the audit. This information can be shared with other auditees so that they can avoid pitfalls if any are encountered elsewhere. The control center should be staffed by people who are ISO knowledgeable and can quickly address potential disasters. When a potential show stopper (a major noncompliance) is encountered, the control center will go into damage control mode and see if they can collect evidence that can downgrade the noncompliance to the "minor" category. This is not done as an attempt to fool the auditors, but is done to insure that all of the evidence possible is brought to their attention. It is often the case that not everyone being interviewed has all of the necessary knowledge concerning an issue.

All escorts and hosts should take numerous notes so that all of the auditors' thoughts are captured, whether or not they turn into noncompliances. After all, an experienced auditor can give valuable insight into what is needed to make the quality system more effective, even though the auditor is not allowed to consult.

If not already scheduled, the auditee's management should request a daily review of findings from the auditors. This will also allow the auditees the ability to do further damage control if the results are not good. The auditors will see through any attempt by the auditees to whip together a missing procedure, so this should not be done. However, if quality records can somehow show that a missing ingredient in the quality system is actually present, then it is valid to bring it to the auditor's attention. In other words, it is permissible to attempt to find missing information but it is not permissible to fabricate it. The auditors are professionals and can detect fabrications and falsifications.

In summary, the implementation plan must be well thought out and rigorously adhered to. It is not a trivial undertaking, so the personnel put in charge of it should have good project management skills. The actual registration audit must be carefully orchestrated to insure that nothing is left to chance and that the auditors will have an opportunity to fairly judge your quality system.

Chapter Review Questions

1. What is a baseline audit and what is its function?
2. What is the advantage of going outside the company for the baseline audit?
3. What are the items to look for when choosing baseline auditors?
4. How does the baseline lead to the project plan?
5. Why must a company take a systemic approach for corrective actions?
6. What is a Gantt chart, and how might it be used during a firm's ISO efforts?
7. What is the role of the ISO coordinator?
8. Discuss ways to prevent the ISO effort from languishing.
9. How can it be determined if the efforts expended as a part of the project plan actually conform to ISO requirements?
10. What is the purpose of the registration audit and who conducts it?
11. What is a typical lead time for the registration audit?
12. Is the registration audit sometimes preceded by a preliminary audit? Why?
13. What are some things to remember about the care and feeding of auditors?
14. What is the purpose of an audit schedule? Who prepares it?
15. Can the registration auditors examine areas not on the audit schedule?
16. What is an audit trail?
17. Who should escort the auditors around the plant?
18. If an auditor asks an operator a question, should the escort answer it first? Is it ever permissible for the escort to assist in the answer?
19. What is an audit control center and who runs it?
20. What is the major purpose of the audit control center?
21. Is it a good practice to ask the auditors for a daily review? Why?
22. In the case of a major nonconformance, is it an acceptable practice to try to quickly formulate and document a missing procedure in order to placate the auditors? Why?

10

REGISTRATION MAINTENANCE AND THE FUTURE

INTRODUCTION

The job is not finished once ISO 9000 registration has been achieved. One of the basic concepts of ISO registration is to insure that once registered, companies can not backslide into old habits without punitive repercussions. It is even worse to have an ISO registration removed from a company than not to have had it in the first place. The ISO 9000 registration process keeps the registered firms honest with the use of surveillance audits. The surveillance audits not only look for maintenance of the current quality system, but also check for system improvement through the use of internal audits, corrective actions, and vigorous management reviews.

SURVEILLANCE AUDITS

Most registration firms will return for surveillance audits at approximately six month intervals. When they return, the auditors will look for some key areas to review. These are:

- weak areas in previous audits
- internal audits
- corrective actions
- management review

Weak Areas

In the registration audit, rarely do companies receive absolutely no noncompliances. Frequently there are a number of system weaknesses noted in ongoing or minor noncompliances or verbal observations that may be noted in the formal written report. These areas will be reviewed as the first order of business of the next surveillance audit. The auditors will expect to see very positive corrective action on the written ongoing noncompliances and some evidence that there has been some improvement activity involving their concern as noted in the observations. If there is a weak response to these noncompliances, it is an indication to the auditors that the firm is not really serious about ISO 9000 and has only gone through the exercise to get their ISO credentials. If the registered firms do not show a very positive system to resolve the nonconformances, the registration firm will begin the process of decertifying the company.

INTERNAL AUDITS

The internal audit system as described in Chapter 5 is an area sure to be closely scrutinized in a surveillance audit. The internal audit system not only serves to demonstrate to management and third party auditors that the quality system remains effective, but it is also the major driving force for continuous improvement in the quality system. The very concept of continuous improvement requires a vehicle which will look for opportunities to make things better. The internal audit system acts as that vehicle for an ISO 9000 compliant quality system.

Corrective Actions

The auditors will review the results of each audit and look for evidence that corrective actions to noncompliances are being aggressively pursued. As with all corrective actions, the auditors will check them to insure that corrective action implementation and effectiveness have been verified. There must be objective evidence in the form of data and analysis of the data to insure effectiveness. The appropriate documents must evidence an authorized signature of someone with the authority to verify. Any corrective actions that have not been completed in a reasonable time period will be evidence that the system may not be working as intended.

The auditors will look at the other elements of the corrective action system as well. The organization must be prepared to show evidence that it is working diligently with its suppliers and using its corrective action system to encourage them to improve their offerings. Weak corrective action responses from suppliers that have been accepted by an organization are an indication that it is not serious enough about controlling suppliers. As stated in Chapter 5, Section 4.6, customers will not receive a quality offering if the process is started with substandard raw materials and services from suppliers.

Problems from internal processes that manifest themselves in customer complaints and internal nonconformances will also be reviewed in surveillance audits. It can not be emphasized enough that the proper processing of corrective actions is the key to the success of a quality system. If a corrective action system becomes ineffective, the quality system will not serve an organization's needs in the global economy, but will wither and die, leaving it extremely vulnerable to competitive erosion of its market position. Registration auditors recognize this and will likely pursue decertification if they see evidence of lack of corrective action resolve.

Management Reviews

ISO 9000 auditors recognize that management involvement is the key to continued success and growth of the quality system. For that reason, they will look for evidence that top management is involved in the process of regularly reviewing the quality system. The auditors will not only look to see that management reviews are happening at the frequency required, but will look at the attendance records to insure that the top manager and key members of staff are present at most reviews.

If the auditors see that key members of management delegate attendance at management reviews to subordinates, they will correctly assume that management is losing interest in nurturing the quality system. This can be the death knell for the maintenance and growth of the quality system. The auditors will also review minutes of the management reviews to insure that action items are being assigned and vigorously pursued. If the auditors see evidence that this is languishing and only being paid lip service, they will conclude that the management reviews are nothing more than an exercise to make it look like management is serious about quality.

Weaknesses in any of the key areas could result in a major noncompliance that would have to be addressed in order to prevent the removal of the ISO 9000 registration. It is unlikely that the first incidence of problems would result in decertification, but it would surely be the first step in the ultimate removal of registered status.

Other Areas

The key areas mentioned above will most certainly be audited in surveillance audits, but that does not mean that these will be the only items of concern. Over time, all areas of the quality system will be audited in a surveillance audit. Usually this happens in a three year cycle. Those areas that may have been a bit weak in the registration audit will be the first re-audited, but all will eventually get the privilege of being audited. Everything is legal game in a surveillance audit, so even though a certain area may not be on the audit schedule, it could still be evaluated if an audit trail leads there. Therefore, it is prudent from an audit point of view as well as for the general well-being of the quality system to keep all aspects of the quality system in good shape.

MAINTENANCE AUDITS

Even though the internal audits should be helpful in keeping the quality system in shape, it is advisable to have an independent third-party audit every once in a while to provide a reality check. These audits can be very helpful even if they find nothing different from those performed internally because they validate the efforts of the in-house personnel. The frequency of these audits is a matter of choice, but the most common frequency is once per year.

THE FUTURE OF STANDARDS

The future will always be a question mark. However, trend forecasting can be acceptably accurate—especially when confined to a reasonable time span. Based on this premise, it is possible to predict that quality standards can only become more pervasive and more important as time goes on.

ISO 9000 has been criticized as being "no guarantee of quality." This statement can be supported to a limited extent, but under scrutiny, it is clear that this statement does not serve any useful purpose beyond being ammunition for detractors and foot-draggers. A fairly obvious rejoinder to this is that no sane company would spend the time and money to work so hard to structure itself and get registered, only to languish or perhaps self-destruct in a more organized fashion. Organizations can languish or self-destruct with much less effort! In other words, ISO lends structure and recognition to a total quality system that should ultimately increase the profitability and health of the company.

The future of ISO 9000 will show an inspired and judicious blending of its tenets and themes with new knowledge and research, and with advanced ideas of total quality management, operational efficiency, financial stability, market growth, and profitability. Detractors may claim that the ISO standards are mutually exclusive with some or all of these. But as time passes the detractors are vanishing because they are being overwhelmed by a global movement toward world-wide standards.

Individual companies are learning how to *benefit* from the ISO standards. It has long been accepted that companies have to invest in the future to ensure their share of the future. Establishing standards and a codified, certified TQM system are among the very best investments that any company can make.

This book has attempted to push the envelope a bit by reporting how international standards can be applied in areas other than manufacturing. It absolutely does not shortchange any critical manufacturing concepts yet it embraces software and service. This is because the authors are of the opinion that standards are vitally needed in many areas of human endeavor. Based on current trends, the future of standards is one of increasing scope. These emerging standards and their applications do *not* have to be onerous, odious, stifling, or oppressive. Standards are opportunities to do important things better!

The core philosophy of future developments will produce carefully structured standards that are not so much prescriptive as they are conceptual. Stated simply, they will help people to do quality work in a consistent fashion, foster ongoing improvements, and accommodate individual differences and creativity. Of course, this will require intelligent and prudent efforts to be realized at the international level. Are these efforts forthcoming? The answer is yes, and as the world continues to shrink it is indubitable that other courses are not in the best interests of people everywhere.

Schools, government agencies, and farms are looking at ISO 9000. It's an idea whose time has come because everyone is a consumer. Consumers deserve the best.

Chapter Review Questions

1. What is the purpose of surveillance audits?
2. How often are surveillance audits conducted?
3. What do the surveillance auditors look for?
4. Who conducts the surveillance audits?
5. What is decertification? If it happens, is it worse than never having attempted to achieve ISO status? Why?
6. Why will surveillance auditors be interested in an organization's internal audit procedure?

7. Can a surveillance audit examine an organization's dealings with its subcontractors? Why?

8. Can a surveillance audit examine an organization's internally generated nonconformances and/or its customer complaints? Why?

9. Why is lack of corrective action one of the most serious deficiencies as far as the surveillance audit is concerned?

10. Why do the surveillance auditors look at the management review meeting schedules, and the meeting minutes including who has attended the meetings?

11. Will all areas of the quality system be eventually reviewed by surveillance audits? About how long would this usually take?

12. Which items and areas stand out as prime fodder for the next surveillance audit?

13. Are there such things as audit trails during surveillance audits?

APPENDIX A

BIBLIOGRAPHY

Alsup, Fred, and Ricky M. Watson. 1993. *Practical Statistical Process Control.* (New York: Van Nostrand Reinhold.)

Besterfield, Dale H. 1994. *Quality Control,* 4th ed. (Englewood Cliffs, NJ: Prentice Hall.)

Carlsen, Robert D., Jo Ann Gerber, and James F. McHugh. 1992. *Manual of Quality Assurance Procedures and Forms.* (Englewood Cliffs, NJ: Prentice Hall.)

Cottman, Ronald J. 1993. *A Guidebook to ISO 9000 and ANSI/ASQC Q90.* (Milwaukee: ASQC Quality Press.)

Crosby, Philip B. 1992. *Completeness: Quality for the 21st Century.* (New York: Penguin Books USA, Inc.)

Fellers, Gary. 1992. *The Deming Vision.* (Milwaukee: ASQC Quality Press.)

Freedman, Daniel P., and Gerald M. Weinberg. 1990. *Handbook of Walkthroughs, Inspections, and Technical Reviews: Evaluating Programs, Projects, and Products,* 3rd ed. (New York: Dorset House Publishing Co.)

Goetsch, David L., and Stanley Davis. 1994. *Introduction to Total Quality.* (New York: Macmillan College Publishing Co.)

Heizer, Jay, and Barry Render. 1993. *Production and Operations Management,* 3rd ed. (Needham Heights, MA: Allyn and Bacon.)

Jackson, Harry K. Jr., and Normand L. Frigon. 1994. *Management 2000.* (New York: Van Nostrand Reinhold.)

Lee, Sang. M., and Marc J. Schniederjans. 1994. *Operations Management.* (Boston: Houghton Mifflin Co.)

Meredith, Jack R. 1992. *The Management of Operations,* 4th ed. (New York: John Wiley & Sons, Inc.)

Owen, D. B. 1989. *Beating Your Competition Through Quality.* (New York: Marcel Dekker, Inc.)

Pressman, Roger S. 1982. *Software Engineering: A Practitioner's Approach,* 3rd ed. (New York: McGraw-Hill, Inc.)

Rawlings, Joseph H. III. 1994. *SCM For Network Development Environments.* (New York: McGraw-Hill, Inc.)

Reilly, Norman B. 1994. *Quality: What Makes It Happen?* (New York: Van Nostrand Reinhold.)

Sashkin, Marshall, and Kenneth J. Kiser. 1993. *Putting Total Quality Management to Work.* (San Francisco: Berrett-Koehler Publishers, Inc.)

Schmidt, Warren H., and Jerome P. Finnigan. 1992. *The Race Without a Finish Line.* (San Francisco: Jossey-Bass, Inc.)

Schroeder, Roger G. 1989. *Operations Management,* 3rd ed. (New York: McGraw-Hill, Inc.)

Tenner, Arthur R., and Irving J. Detoro. 1992. *Total Quality Management.* (Redding Heights, MA: Addison-Wesley Publishing Co., Inc.)

Townsend, Patrick L., and Joan E. Gebhardt. 1992. *Quality In Action.* (New York: John Wiley & Sons, Inc.)

Vonderembse, Mark A., and Gregory P. White. 1991. *Operations Management,* 2nd ed. (Saint Paul, MN: West Publishing Co.)

APPENDIX E

QUALITY VOCABULARY
(Adapted From ISO 8402 and 9000-3)

product: A tangible result of activities or services or an intangible such as a design, directions for use, or a service.

service: An activity or process.

3.1 quality: All the features of a product or service that affect its ability to satisfy stated or implied needs. Needs are specified in contractual arrangements; implied needs should be identified, defined, and updated as required. Needs may include usability, safety, availability, reliability, maintainability, economics, and environment. The term "quality" should *not* be used as a measure of comparative excellence. Nor is it a quantitative measure of technical evaluation. The term "relative quality" can be used when products or services are to be comparatively ranked. The terms "quality level" or "quality measure" can be used for quantitative measures. Because quality is influenced by interactive activities, phrases such as "quality attributable to design" and "quality attributable to implementation" can be used to emphasize specific activities.

3.1 software: Intellectual creation comprising the programs, procedures, rules and any associated documentation pertaining to the operation of a data processing system. Software is independent of the medium on which it is recorded.

3.2 grade: The category or rank related to characteristics that cover different sets of needs for products or services intended for the same functional use. Grade reflects a planned or recognized difference in requirements with an emphasis on the functional use to cost ratio. A high grade item can be of inadequate quality for satisfying needs.

3.2 software product: Complete set of computer programs, procedures, and associated documentation and data designated for delivery to a user.

3.3 quality loop or spiral: The conceptual model of interacting activities that influence quality ranging from the identification of needs to the assessment of needs satisfaction.

3.3 software item: Any identifiable part of a software product at an intermediate step or at the final step of development.

3.4 quality policy: The formal expression by top management pertaining to an organization's intentions and direction regarding quality.

3.4 development: All activities to be carried out to create a software product.

3.5 quality management: That aspect of the overall management function that determines and implements the quality policy. Quality management includes strategic planning, allocation of resources, quality planning, operations, and evaluations. The attainment of quality requires the commitment and support of all members of an organization but the responsibility for quality management belongs to top management.

3.5 phase: Defined segment of work. A phase does not imply the use of any specific life-cycle model, nor does it imply a period of time in the development of a software product.

3.6 quality assurance: All systematic actions required to provide confidence that a product or service will satisfy the quality requirements which must fully reflect user needs. Providing confidence may involve producing evidence via production audits, verification of installations, and inspections. Quality assurance is a management tool and also provides customer confidence.

3.6 verification: For software, the process of evaluating the products of a given phase to ensure correctness and consistency with respect to the products and standards provided as input to that phase.

3.7 quality control: The operational activities and techniques used to fulfill the requirements for quality that are involved with process monitoring and the elimination of the causes of unsatisfactory results. Modified terms such as "manufacturing quality control" or "corporate quality control" should be used to enhance specificity and avoid confusion.

3.7 validation: For software, the process of evaluating the software to ensure compliance with specified requirements.

3.8 quality system: The organizational structure which includes the procedures, responsibilities, processes, and the resources for implementing quality management. The system should be only as comprehensive as needed to meet the quality objectives. Elements of the system may require demonstration to meet contractual and assessment requirements.

3.9 quality plan: Documented practices, resources, and activities that are relevant to a particular product, service, contract, or project.

3.10 quality audit: A systematic and independent examination to determine if activities and results comply with the plan and if they are properly implemented and suitable. The audits are typically applied to a system or its elements and are called "quality system audit," "process quality audit," or "service quality audit." The audits are conducted by staff with no direct responsibility in the areas being audited but should have the cooperation of relevant personnel. Audits determine the need for improvement or corrective action and should not be confused with "surveillance" or "inspection" activities that are a part of process control or product acceptance. Quality audits can be conducted for internal or external purposes.

3.11 quality surveillance: The on-going monitoring and verification of procedures, methods, conditions, processes, products and services, and the analysis of records to ensure that quality requirements are being met. The surveillance may be carried out by or on behalf of the customer. Surveillance should consider factors which could result in deterioration or degradation with time.

3.12 quality system review: A formal evaluation by top management of the status and adequacy of the quality system in relationship to the quality policy and any new objectives resulting from changing circumstances.

3.13 design review: A formal, documented, comprehensive and systematic examination to evaluate requirements and the capability of a design to meet those requirements, to identify problems, and to propose solutions. The review is not intended, by itself, to ensure proper design and it can be conducted at any stage of the design process. The review examines fitness for purpose, feasibility, manufacturability, measurability, performance, reliability, maintainability, safety, environmental issues, time scale, and life cycle costs. The review should be conducted by staff representing all functions affecting quality.

3.14 inspection: Activities such as measuring, examining, testing, and gauging one or more characteristics and comparing these with the specified requirements to determine conformity.

3.15 traceability: The ability to trace the history, application, or location of an item or activity, or similar items or activities, by means of records. There are three main areas: (1) in the distribution sense, it relates to a product or service; (2) in the calibration sense, it relates to measuring equipment, standards, physical constants, or properties; (3) in the data sense, it relates to calculations and data generated throughout the quality loop of a product or service. The records should be specified for a stated period of history or to a point of origin.

3.16 concession (waiver): A written authorization to use or release material already produced that does not conform to specified requirements. These should be for limited quantities or periods and for specified uses.

3.17 production
permit (deviation
permit): A written authorization, prior to production or before the provision of a service, to depart from the specified requirements for a specific quantity or a specific period of time.

3.18 reliability: The ability of an item to perform a function under stated conditions for a stated period of time. Also a characteristic that denotes a probability of success.

3.19 product
liability (service
liability): The legal responsibility of a producer or provider to make restitution for loss related to injury, damage, or other harm caused by a product or service. Liabilities are controlled by local, regional, or national laws.

3.20 nonconformity: The nonfulfillment of specified requirements including the departure or absence of one or more specified quality system elements.

3.21 defect: The nonfulfillment of intended usage requirements including the departure or absence of one or more quality characteristics from the intended usage requirements.

3.22 specification: The documentation that prescribes the requirements with which the product or service must conform. Specifications should refer to or include drawings, patterns, and other relevant documents and should also indicate the means of checking for conformity.

Additional Vocabulary (not in ISO 8402)

accuracy: The degree to which a measure conforms to an accepted standard. Correctness. Not to be confused with precision.

applicant: The manufacturer or supplier making application for registration.

auditor: An external or internal, approved, individual that assays a process, a system, a product, a service, and all relevant documentation to determine compliance with established standards.

benchmark:	A standard of performance.
calibration:	The verification or adjustment of the scales of a measuring instrument against a standard or another calibrated instrument.
capability:	The natural variation of a process due to expected causes.
certification:	The act of documenting compliance with standards and requirements such as products, personnel, processes, and services. Not to be confused with registration.
characteristic:	A quantitative or qualitative measure of a process, product, or service.
conformance:	Meets all specified characteristics.
critical:	A characteristic that significantly affects quality and/or customer satisfaction.
deficiency:	Absence or shortfall of one or more specified characteristics.
instrumentation:	The tools and equipment used in the metrics of process and/or product verification.
metrics:	The theory and practice of measurements.
metrology:	The science of measurements including measurement standards.
obsolete:	No longer relevant or acceptable (for example, obsolete documents must be removed from all locations).
parameter:	A measurable characteristic of a product or process.
precision:	The resolution/repeatability of a process or measurement. Not to be confused with accuracy (an instrument might consistently yield 4 significant digits, but be uncalibrated and thus not accurate).
proficiency:	The ability to perform or produce in adherence to established standards.
purchaser:	The recipient of products and/or services delivered by the supplier.
qualification:	The process of demonstrating if an entity is capable of meeting specified requirements.

registration:	Proof of compliance, as verified with an audit by an accredited external body using a standard such as ISO 9001. Not to be confused with certification.
standard:	A public, documented, and organized body of specifications and features, authored and/or agreed to by the groups affected by it.
statistics:	The mathematical science of sampling and variability.
subcontractor:	An organization which provides product and/or service to the supplier.
supplier:	The organization to which the ISO standards apply.
qualification:	The determination if a procedure, process, product or service meets established minimum standards.
workmanship:	The quality of the work produced by employees.

Appendix C

Common Acronyms

AALA	American Association for Laboratory Accreditation
AFIT	Air Force Institute of Technology
AIA	Aerospace Industries Association
AMC	Army Material Command
ANSI	American National Standards Institute
ANOVA	Analysis of Variance (SPC tool)
ANSI	American National Standards Institute
AQAPS	Allied Quality Assurance Publications
AQL	Acceptance Quality Level (or Limit)
ARL	Average Run Length (SPC tool)
ASME	American Society of Mechanical Engineers
ASQC	American Society for Quality Control
ASTM	American Society for Testing and Materials
ATQMC	Army Total Quality Management Committee
BIPM	International Bureau of Weights and Measures
BOM	Bill Of Materials
BSI	British Standards Institution
CAD	Computer Assisted Design (or Drawing)
CAE	Computer Assisted Engineering
CAM	Computer Assisted Manufacturing
CASCO	Conformity Assessment Standards Committee (of ISO)
CASE	Coordinating Agency for Supplier Evaluation (also Computer Assisted Software Engineering)
CAT	Corrective Action Team

CD	Critical Dimension
CDA	Commercial Distributors Auditing (program)
CDRL	Contract Data Requirements List
CEN	Committee for European Standardization (equivalent to ISO)
CENELEC	Committee for European Electrotechnical Standardization (equivalent to IEC)
CEO	Chief Executive Officer
CFE	Contractor Furnished Equipment
CIM	Computer Integrated Manufacturing
CNC	Computer Numerical Control (automated manufacturing equipment)
COC	Chamber of Commerce
COMPASS	Compliance Management and Product Assurance
CPM	Critical Path Method
CRAG	Contractor Risk Assessment Guide
CRP	Capacity Requirements Planning
CRRL	Certified Record of Released Lots
CSA	Canadian Standards Association
DAR	Defense Acquisition Regulation
DOC	Department of Commerce
DOD	Department of Defense
DODISS	Department of Defense Index of Specifications and Standards
DPSO	Defense Product Standards Office
DTI	Department of Trade and Industry (United Kingdom)
EC	European Community (Belgium, Denmark, France, Ireland, Italy, Luxembourg, The Netherlands, Portugal, Spain, United Kingdom, Germany)
EEC	European Economic Community
EFTA	European Free Trade Association (Austria, Finland, Iceland, Norway, Sweden, Switzerland)
EIA	Electronic Industries Association
EOQ	Economic Order Quantity

EOQC	European Organization for Quality Control
EPQ	Economic Production Quantity
EQS	European Committee for Quality System Assessment and Certification
ETSI	European Telecommunications Standards Institute
EVOP	Evolutionary Operations (SPC tool)
FAA	Federal Aviation Administration
FAR	Federal Acquisition Regulation
FMS	Flexible Manufacturing System
FTC	Federal Trade Commission
GAO	General Accounting Office
GATT	General Agreement on Tariffs and Trade
GFE	Government Furnished Equipment
GIDEP	Government Industry Data Exchange Program
GNP	Gross National Product
GSA	General Services Administration
IEC	International Electrotechnical Commission
IECQ	International Product Certification System for Electronic Products
IEEE	Institute of Electrical and Electronic Engineers
IG	Inspector General
IROR	Internal Rate of Return
ISCPC	International Standards and Certification Policy Committee
ISO	International Organization for Standards
JAN	Joint Army and Navy
JDS	Job Diagnostic Survey
JEDEC	Joint Electronic Device Engineering Council
JIS	Japanese Industrial Standards
JIT	Just In Time (inventory and manufacturing control system)
JMA	Japan Management Association

JMI	Japanese Manufacturer Institute
JUSE	Japanese Union of Scientists and Engineers
LCL	Lower Control Limit (SPC tool)
LRQA	Lloyds Register Quality Assurance (United Kingdom)
LTPD	Lot Tolerance Percent Defective
MAD	Mean Absolute Deviation (SPC tool)
MBWA	Management By Walking Around
MITI	Ministry of International Trade and Industry (Japan)
MOU	Memorandum Of Understanding
MPCAG	Military Parts Control Advisory Group
MPS	Master Production Schedule
MRB	Material Review Board
MRP	Material Requirements Planning (also Manufacturing Resource Planning)
MSE	Mean Squared Error (SPC tool)
MTBF	Mean Time Between Failures
MTM	Methods Time Measurement
NACCB	National Accreditation Council for Certification Bodies (United Kingdom)
NASA	National Aeronautics and Space Administration
NATO	North Atlantic Treaty Organization
NC	Numerical Control
NCAS	National Contractors Accreditation System
NCSL	National Conference of Standards Laboratories
NIH	Not Invented Here (a myopic point of view)
NISO	National Information Standards Organization
NIST	National Institute of Standards and Technology (was NBS)
NSC	National Safety Council
OEM	Original Equipment Manufacturer
OFPP	Office of Federal Procurement Policy

OIML	International Organization of Legal Metrology
OPT	Optimized Production Technology
OSTT	Office of Strategic Trade and Technology
PC	Process Control
PCS	Project Control System
PDA	Percent Defective Allowable
PDM	Precedence Diagramming Method (CRP tool)
PFA	Product Flow Analysis
PM	Process Monitor
Q&R	Quality and Reliability
QA	Quality Assurance
QC	Quality Control (or Circle)
QCI	Quality Conformance Inspection
QFD	Quality Function Deployment
QIT	Quality Improvement Team
QML	Qualified Manufacturer List
QMS	Quality Management System
QPL	Qualified Products List
QRE	Quality and Reliability Engineering
QSR	Quality Systems Registrar, Inc.
QST	Quality Steering Team
QVL	Qualified Vendors List
RAB	Registrar Accreditation Board
RAC	Reliability Analysis Center
RAPID	Receiving Assessment and Parts Identification
RE	Reverse Engineering
RvC	Raad voor de Certificatie (The Netherlands)
SA	Standards of Australia
SDP	Software Development Plan
SEI	Software Engineering Institute

SFPE	Society of Fire Prevention Engineers
SII	Standards Institute of Israel
SOP	Standard Operating Procedure
SPC	Statistical Process Control
SQSR	Software Quality System Registration
TAG	Technical Advisory Group
TELARC	New Zealand National Accreditation Authority
TQC	Total Quality Control
TQM	Total Quality Management
UCL	Upper Control Limit (SPC tool)
UL	Underwriters Laboratories
WIP	Work In Progress

APPENDIX D

FORMS AND CHECKLISTS

Approved Commodity and Supplier List

Commodities (list in alphabetical order)	Approved Supplier	Rating (1, 2, or 3)

Rating System: 1 = Preferred, 2 = Adequate, 3 = Only in emergencies

| Approved Supplier List | | |

Name of Company	Commodities Purchased	Rating (1, 2, or 3)

Rating System: 1 = Preferred, 2 = Adequate, 3 = Only in emergencies

Authorization to Release Material
for Urgent Production Needs

I authorize the material described below to be used for immediate production without first being verified for the proper quality.

Samples of this material or other evidence will be provided at a later date. If a nonconformance is detected, the material and any product processed with the material may be recalled.

Signed _____ Date _____

Description of Material:

4

CALIBRATION MASTER INVENTORY

Form No. [   ]

Equipment or Instrument	Serial No.	Last Date	Due Date	Location	Owner

Sheet _____ of _____

Calibration Cycle Start Date _____ End Date _____

Information Supplied By _____

Approved by _____

CALIBRATION/SERVICE RECORD	Serial No. _____ Recalibrate Due Date
Item Name Model	Manufacturer

Measurement Standards Required

Repair History	Date

CALIBRATION DATA

Parameter	Nominal Value	Measured Value	Corrected Value	Minimum Limits	Maximum Limits

Notes and Remarks

Signature _____ Date _____

CONTRACT QUALITY REQUIREMENTS ANALYSIS

Form No. [ ]

Customer	Contract No.	Contract Type

Product Description

Start Date	End Date	Quantity	Max. Rate

Similar to previous contracts?　☐ No　☐ Similar　☐ Identical
If similar or identical, explain.

Are Quality Terms Stated?　☐ Yes　☐ Difficult to Interpret　☐ No

---

Summary Statement

Advanced Metrology Required

Special Equipment/Fixtures/Instruments/Tools

Special Skills and Training Required

Workload/Employees Required

Other

Analysis By	Date
Reviewed By	Date

CONTROLLED DOCUMENT INVENTORY

Form No.

Form Revision Date _____   Sheet _____ of _____

DOCUMENT NAME	DOCUMENT NUMBER	REVISION DATE	LOCATION	OWNER

Information Supplied By _____

Document Control Administrator _____

Corrective Action Request Form

Date_____   Corrective Action Request No. _____

Part Name _____   Part No. _____   Lot No. _____

Description of Problem or Nonconformance:

Root Cause Analysis:

Corrective Action Plan:

Signature_____   Date _____   Approved _____

Implementation Verified _____ Date _____ Approved _____

Effectiveness Verified _____ Date _____ Approved _____

DESIGN PROTOCOL PHASE 2 CHECKLIST

Project Name _____   Project No. _____

Task	Timing	Responsibility
Up-to-date customer prints		
Customer application defined		
Material specification		
Patent application filed		
Reliability analysis		
Packaging specification		
Process flow diagram		
Preliminary mill cost estimates		
Preliminary staffing estimate		
Preliminary quality plan		

Signature	Date
Engineering	
Manufacturing	
Quality	
Marketing	

DESIGN QUALITY REVIEW

Form No.

Project Title		Project No.	
Item	Item No.	Review No.	Date
Location		Design Cycle Phase	
Name of Attendees		Departments/Organizations	

Quality Topics Discussed

Action Items Completed from Last Review ——————————————

Action Items Overdue ——————————————————————

Design Details Sheets Attached? ☐ Yes   Number _____ ☐ No

Action Required	Assigned To	Complete By Date

Prepared By	Date
Approved By	Date

DOCUMENT CHANGE ALERT	Form No.
Document Name	Document No.
Area/Department	Date
Nature of Change	

We the undersigned have read the revised document and understand the changes as described above.

Name	Department	Date

INTERNAL AUDIT REPORT

Form No.

Location	Reference No.
Assessor's Signature	Date

System Elements / Departments / Findings

Name and Number of Noncompliance Cleared
From Previous Audit

Name and Number of Uncleared Noncompliance

Name and Number of New Noncompliances

Next Audit Due

Special Remarks

INTERNAL AUDIT SCHEDULE

Form No.

Location	Submitted By
Approved By	Date
Special Notes	

Element to be Examined	Dates			
Management Responsibility				
Quality System				
Contract Review				
Design Control				
Document Control				
Purchasing				
Purchaser Supplied Product				
Identification and Traceability				
Process Control				
Inspection and Testing				
Inspection Equipment				
Inspection Software				
Equipment Calibration				
Noncomforming Product				
Corrective Action				
Handling and Protection				
Quality Records				
Internal Audits				
Training				
Servicing				
Statistical Techniques				

ISO 9001 REVIEW CHECKLIST Form No.

Subject	Element	Responsibility	Date	Status
Management Policy	4.1.1			
Responsibility & Authority	4.1.2.1			
Verification Resources & Personnel	4.1.2.2			
Management Representative	4.1.2.3			
Management Review	4.1.3			
Quality System	4.2			
Contract Review	4.3			
Design Control—General	4.4.1			
Design & Development Planning	4.4.2			
Activity Assignment	4.4.2.1			
Organizational & Technical Interfaces	4.4.2.2			
Design Input	4.4.3			
Design Output	4.4.4			
Design Verification	4.4.5			
Design Changes	4.4.6			
Document Approval & Issue	4.5.1			
Document Changes/Modifications	4.5.2			
Purchasing—General	4.6.1			
Assessment of Sub-Contractors	4.6.2			
Purchasing Data	4.6.3			
Verification of Purchased Product	4.6.4			
Purchaser Supplied Product	4.7			
Product Identification & Traceability	4.8			
Process Control—General	4.9.1			
Special Processes	4.9.2			
Inspection & Testing—Receiving	4.10.1			
In Process Inspection & Testing	4.10.2			
Final Inspection & Testing	4.10.3			
Inspection & Test Records	4.10.4			
Inspection, Measuring & Test Equipment	4.11			
Inspection & Test Status	4.12			
Control of Nonconforming Product	4.13			
Nonconformity Review & Disposition	4.13.1			
Corrective Action	4.14			
Handling, Storage, Packaging—General	4.15.1			
Handling	4.15.2			
Storage	4.15.3			
Packaging	4.15.4			
Delivery	4.15.5			
Quality Records	4.16			
Internal Quality Audits	4.17			
Training	4.18			
Servicing	4.19			
Statistical Techniques	4.20			

NOTIFICATION OF NONCOMPLIANCE

Form No.

Location	Reference No.	Date
Element	Specification	Note Number

Assessor's Name	Recipient's Name
Date of Issue	Date of Receipt

Description of the Noncompliance and Related Conditions

Location's Planned Corrective Action(s)

Planning Date_____

Approved by: Signature _____

Verification of Corrective Action(s)

Action Implemented   Signed _____ Date _____

Action Effective   Signed _____ Date _____

Notification of Software Contract Review

Customer Name _____

Project Name _____

Project Number _____

Document Identifier _____

Contract Name _____ Number _____

Other Referenced Documents _____

Date _____ Time _____

Location _____

Participant: Supplier _____

                       _____

                       _____

            Purchaser _____

                       _____

                       _____

Notes and Exceptions:

Approved By _____ Date _____

Software Contract Review Report

Customer Name _____   Date _____

Project Name _____   Project No. _____

Other Referenced Documents _____

Participants: Supplier _____

_____

_____

Purchaser _____

_____

_____

Outstanding Items Needing Clarification

Requirement _____ Page No._____

Requirement _____ Page No._____

Requirement _____ Page No._____

Requirement _____ Page No._____

Requirement _____ Page No._____

Terminology Conflicts _____

_____

Technology Conflicts_____

_____

Action Items _____

_____

Approved By: Supplier _____ Date _____

                    Purchaser_____ Date _____

Software Inspection and Formal Review Error List

Customer _____

Project Name _____

Software Version No. _____

Date of Inspection or Review _____

Page _____ of _____

Error No. _____ Error Type _____

Page No. or Line No. _____ Noted By _____

Enter Detailed Error Description:

Priority: ☐ High ☐ Medium ☐ Low ☐ Other _____

Special Notes _____

Hardware _____

Operating System _____

Special Conditions _____

Consistent Problems _____

Intermittent Problems _____

Signature _____ Date _____

Software Inspection and Formal Review Summary

Customer _____ Project _____ Date of Review _____

Starting Time _____ Ending Time _____ Software Version No. _____

Process, Unit, or Function Name _____ Department _____

☐ Code Inspection ☐ Design or Document Review ☐ Other _____

Type of Review: ☐ Requirement Specification ☐ Code

☐ Functional Specification ☐ Design Specification

☐ Operator Documentation ☐ Test Plan

Additional Materials Used for Review:

Reinspection _____

Preparation Time (in hours)_____ Duration of Inspection _____

For Code Inspections, Total Lines of Source Code _____

Estimated Hours for Revisions and Corrections _____

If Document, Total No. of Lines ____ If Reinspection, Scheduled Date _____

Review Participants

Author _____ Phone _____

Author _____ Phone _____

Leader _____ Phone _____

Recorder _____

Review Team _____

Result of Review and Inspection: ☐ Review Not Completed

☐ Accepted (no reinspection required) ☐ Not Accepted (reinspection required)

☐ With Minor revisions ☐ With Major Revisions

See Related Documents:
Error List _____
Related Issues List _____
Other _____

Software Maintenance Report

Customer Name _____

Project Name_____

Software Version _____

Type of Maintenance Requested _____

Enhancement or Modification _____

Bug Fix _____

Description of Enhancement or Modification _____

_____

_____

Description of Bug _____

_____

_____

Hardware Configuration _____

_____

Submitted By_____ Date Submitted _____

Request of Report_____

Approved By _____

Subcontractor Quality Assessment Summary	Form No.

Company Name	General Product Line
Address	Total No. of Employees
Phone	No. of Production Employees
Contact Person/Department	No. of Quality Assurance Employees

Potential items from this source:

Quality specifications compatible with this source:

Special areas (temp-humid control, clean room, etc.):

Is their inspection and test equipment traceable to N.I.S.T.

☐ Yes   ☐ No   Comment:

Their quality system includes:

☐ ISO 9000 registration   ☐ quality department   ☐ process inspection
☐ external audits   ☐ training   ☐ records/reports
☐ internal audits   ☐ corrective action   ☐ incoming inspection
☐ quality manual   ☐ storage control   ☐ design control
☐ inspection stamps   ☐ final inspection   ☐ statistical methods
☐ document control   ☐ calibration   ☐ purchase control

Special expertise or capabilities:

Notes and remarks:

Attached copies of their QA charts/procedures ☐ Yes ☐ Pages ☐ No	
Prepared by: Date:	Approved by: Date:

SUPPLIER RATINGS			
Supplier	No. of Deliveries	No. of Problems	Percentage
Aztrone Metals	145	9	94
Okra Chemical	78	0	100

TEMPORARY CHANGE FORM   Form No.

Area	Date
Initiated By	
Part Number	
Drawings Affected	
Procedures Affected	

Description of Change

Reason for Change

Expiration Date

Authorized By   Date

TEST DATA RECORD

Part _____ Test _____

Project _____ Specification _____

High Value Limit _____ Low Value Limit _____

Date	Time	Measured Value	Employee

Training Matrix

Department _____

Name	Operate Machine	Setup Machine	Perform Maint.	Inspect Product	Package Product	Final Audit

Training Request Form

Employee Name _____ S.S. No. _____

Department _____ Date _____

Course Name _____ Course Date _____

Course Length _____ Course Cost _____

Purpose of Course:

Approvals _____

_____

_____

VENDOR CORRECTIVE ACTION REQUEST

Form No. _____

Request No.	Date

TO:

FROM:

P.O. No.	Part No.	Part Name
Quantity Received	Quantity Rejected	Quantity Returned

Description of Discrepancy

Disposition

☐ Returned for Evaluation   ☐ Returned for Rework

☐ Used As Is   ☐ Reworked at Your Expense

☐ Other _____   ☐ Other _____

_____   _____
Purchasing Approval   Quality Assurance Approval

VENDOR TO COMPLETE THE FOLLOWING

Cause of Discrepancy

Corrective Action Taken to Eliminate Recurrence

Date Implemented _____ Date Verified as Effective _____

Approval Signature _____ Title _____ Date _____

APPENDIX E

STAMPS AND TAGS

STAMP	NAME	USE
(A 13)	Source Material Inspection	To certify that incoming material or item has been inspected and accepted.
(A 19)	In-Process Inspection	To certify that partially processed material or item has been inspected and accepted.
A⊤18	Test Acceptance	To certify that item has been tested and found functionally acceptable.
[A 23]	Final Inspection	To certify that processed material or item has been inspected and accepted.
[7]	Discrepant Material	To identify material as discrepant or non-conforming.
/A 14\	Material Review Acceptance	To identify material accepted by the material review board.
℞2	Material Review Rejection	To identify material rejected by the material review board.
QE 17	Quality Engineering	Certifies approval by quality engineering.
(CAL 15)	Calibration	To identify instruments and equipment that have been calibrated to standards.

1. Each stamp has a unique number and is assigned to an individual.

2. The stamp imprint and the signature of the owner is recorded.

3. Worn or obsolete stamps are returned to department of issue.

4. Stamps returned due to personnel changes are not reissued for one year.

5. Written notification of a lost stamp is required.

6. The unique number of a lost stamp will not be reused for two years.

7. A stamp is declared invalid by written notification to all parties.

8. All stamps will be subject to periodic audit.

9. An accurate and up-to-date log of all stamps is required.

SOURCE INSPECTION

☐ Final ☐ Partial No. _____

Vendor _____ Date _____

Part Name _____

Part No. _____ P.O. No. _____

Quantity Accepted _____ Quantity Rejected _____

Comments _____

_____

_____

| Inspection |
| Stamp |

TEST ACCEPTANCE

Work Order or P.O. _____ Date _____

Part Name _____ Part No. _____

Test Name _____

Quantity Passed _____ Quantity Failed _____

Comments _____

_____

_____

| Inspection |
| Stamp |

SCRAP

Part Name _____ Date _____

Part No. _____ Quantity _____

Work Order or P.O. _____

Material Review No. _____

Comments _____

_____

_____

| Inspection |
| Stamp |

LIMITED SHELF LIFE

MUST BE USED BEFORE (DATE) _____

MATERIAL DESCRIPTION _____

SUPPLIER _____

PURCHASE ORDER _____

COMMENTS _____

_____

_____

Inspection

Stamp

STATIC - SENSITIVE

This Material Must Be Worked On
And Stored In A Static-Free Environment

Material Description _____

Supplier _____

Purchase Order _____

Comments _____

_____

_____

Inspection

Stamp

TEMPERATURE - SENSITIVE

Temperature Range To: _____ °C _____ °F

Material Description _____

Supplier _____

Purchase Order _____

Comments _____

_____

_____

Inspection

Stamp

CALIBRATION DUE

Equipment No. ————————

DATE ————————————

This Item Last Calibrated

Date | By (Stamp)

CAL 15

NOT CALIBRATED

NOT TO BE USED
FOR OFFICIAL
MEASUREMENT PURPOSES

OUT OF SERVICE

MUST BE REPAIRED
AND/OR CALIBRATED
BEFORE USE

Name ——————— Date ———

NO CALIBRATION
REQUIRED

Equipment No.

By (Stamp) | QE 17

PART NO. _____

ORDER NO. _____

PROCESS	MACHINE NO.	COMPLETED BY	DATE
INCOMING INSPECTION			
PRETREAT			
ASSEMBLE			
IN-PROCESS INSPECTION			
HEAT TREAT			
PAINT			
INSPECT			
PACKAGE			
FINAL AUDIT			

LOT TRAVELER

Index

A

acceptance, 144
acceptance level, 142
acceptance test plan, 144
accreditation, 19, 35, 37–39, 41
accreditation and registration process, 39
accreditation process, 38
activity based sequential chart, 177
administrative control function, 68
administrative process, 87
adversarial relationship, 36, 118
agenda for a quality review, 49
aggregate planning, 166, 167, 175
aggregate resources, 166
alerts, 59, 71
Allied Quality Assurance Publication 1, 5
American National Standards Institute (ANSI), 38, 140
American Society for Quality Control, 5, 37, 182
appointment systems, 167
approval status, 128
approved suppliers list, 23, 29, 57, 80, 87, 98
aspects of, 126
assess subcontractors, 168
assessment of subcontractors, 154
audit escorts, 183
audit management, 182
audit records, 100
audit results, 88, 91, 93
audit schedule, 91, 182, 184, 188
auditing procedures, 16
   internal quality audits, 48
   management review, 187
   schedule of audits, 91
auditing process, 92, 93
   records, 93
auditors, 8, 25, 29, 33, 39–41, 72, 74, 76, 91–93, 123, 124, 128, 178, 182–87, 189, 190. *See also* consultants
   auditing and communicating, 178
   auditor registration, 40
   grades of RAB certified auditors, 40
   independence, 92
   qualifications for auditors, 39
   training costs, 92
   typical audit, 74
author, 141
automated (computerized) inventory/ production management, 109
automation, 57, 58, 110, 163
available production hours, 166

B

baseline audit, 177, 178, 184
benchmark audit, 177. *See also* baseline audit
benchmarks, 14, 15
bidding process, 17
business investment, 30

C

calibration certificate, 80
calibration information, 76, 80
calibration laboratory, 82
calibration program, 80
   secure environment, 82
calibration records, 59, 76
calibration technician, 81
capacity requirements planning, 166
central documentation area, 68
chain reaction of quality reduction, 17

change request review board, 149
Check IT (Information Technology), 123.
    See also TickIT Program
code inspection, 140, 147
    checklist for, 141
communications, 14, 37, 104, 121, 154,
    171, 172, 174, 178, 180
company image, 19
company organization chart, 50
compression, 167
computer software, 67, 74, 121
    contract requirements, 129
    database-driven applications, 154
    purchaser's responsibility, 125
    quality in software development, 122
concurrent design model, 64
concurrent engineering, 61, 62
configuration management, 8, 126, 132,
    136, 139, 148–50, 153, 157
configuration management activities,
    149
configuration management plan, 150,
    157
configuration status reports, 150
conformance, 5, 19, 144, 179, 184
consultants, 26, 30, 33–37, 42, 53, 154, 181
    ability to communicate, 35
    advantages of hiring, 33–35
    cost of hiring, 33
    experience with other companies, 36
    external degree granting institutions, 35
    ISO 9000 quality system
        implementation experience, 36
    selecting, 34
consumers, 2, 4, 11–15
    customer acceptance, 87
    customer needs and expectations, 12
contingency plans, 164, 166
continuous improvement loop, 62
continuous improvement process, 92
continuous improvement system, 107
contract, 102, 106, 112, 121, 123, 125,
    127, 129, 130, 133, 143–46, 155–57,
    169
contract review, 19, 55, 71, 96, 97, 106,
    112, 127, 129
control center, 183, 184
control chart, 94, 95

control limits, 95, 96
    recalculation, 96
control of customer supplied product, 58
control of inspection, measuring and
    test equipment, 79
control of purchasing, 168
controlled conditions, 73. *See also*
    process control
controlled copy, 69, 75
controlled documents, 51, 56, 67–70, 76,
    97, 108
controlled stamp, 68, 69
copyright and licensing terms, 145
corrected data, 72
correcting any deficiencies, 128
corrective action system, 24, 58, 89, 186,
    187
corrective actions, 20, 38, 55, 89, 90, 93,
    147, 153, 173, 179, 184–87
corrective and preventive action, 88, 90,
    91, 93, 152
    audit records, 93
    follow-up system, 90
    insufficient resources, 89
    "preventing recurrence", 104
cost containment, 90
cost reductions, 16
cross functional contributions, 55
customer audit, 7, 33, 37
customer communication, 172
customer complaints, 12, 46, 79, 88, 89,
    95, 152, 153, 171, 172, 187, 190
customer contact process, 162
customer information system, 155
customer needs and expectations, 12,
    13, 161
customer non-contact processes, 162
customer satisfaction, 19, 46, 48, 55,
    102, 106, 160, 169, 171, 176
customer scheduling tactics, 167
customer surveys, 48, 152
customer training, 145
customer transactions, 56
customer-client relationship, 34

D

damage control, 183
decentralizing purchasing, 168

decertifying, 186
delinquent calibration, 81
delivery, 56, 72, 86, 87, 99, 127, 143–45, 149, 151, 152, 162–65, 169
departmental organizational and flow charts, 50
design and development, 7, 63
design and implementation stages, 138
design constraints, 131
design control, 60, 61
design cycle, 14, 58, 62
  customer needs and satisfaction, 62
  market readiness, 66
  potential problems, 62
  process specifications and service requirements, 65
  protocol, 60
  reviews, 60
  verification, 61
design feedback, 139
design function, 24, 62
design philosophy, 61
design review and corrective action, 13
detractors, 188, 189
developers methods and procedures, 139
development life cycle, 149. See also product life-cycle
development plan, 133–37, 156
  development phases
      required inputs and outputs, 136
      verification activities, 137
  organizational responsibilities, 135
development process, 121, 122, 124, 126, 127, 130, 133–36, 148, 151, 153
disciplined process, 134
document approval and issue, 151
document change, 108
document control, 19, 24, 29, 66, 68–71, 97, 104, 108, 109, 139, 150, 157. See also documentation process
  authorization process, 69
  backup media, 67
  control of computer software, 67
  controlling changes, 70
  disciplined process, 71
  document changes, 151
  guidelines for, 108
  keys to implementation, 68
  method of control, 70
  requirements, 68
  types of documents, 150
document control group, 70, 109
document control personnel, 68–69
documentation process, 28, 69. See also quality manual
  administrative process, 66
  documented quality system, 19
  Level I, II, III, IV documents, 50–51
  manufacturing process, 66
  retrieve company knowledge, 29
documented quality system, 48, 49, 76, 126, 128
  loss of key personnel, 20
duty tours, 167

E

economic ills, 2
economic outlook, 2
electronic interchange of information, 18
electronic lot records, 73
employee attitudes, 106
employee quality handbook, 55
employee schedules, 167
environmental conditions, 76
environmental controllers, 58
ethics, 17, 35
European Union (EU), 4
European Coal and Steel Community (ECSC), 4
European Community (EC), 4, 5, 7, 9, 41. See also European Union (EU)
  member nations, 5
European Free Trade Association (EFTA), 7, 38
evaluating suppliers, 29
evaluation system, 77
exception reporting, 172, 176
excess inventory, 116
executive and management training, 54
expiration dates, 87
exploded, 110
external audit, 8, 33, 177
external degrees, 35

F

fabrications and falsifications, 183
false control limits, 95
feedback information, 12, 14. *See also*
    customer needs and expectations
    monitor the quality characteristics of
    the product, 13
field servicing, 59
field testing, 142, 143
final inspection, 7, 79
final inspection and testing, 7
final product, 77, 86, 87, 116, 130, 131, 145
final product acceptance, 131
flextime, 167
flextour, 167
floating workers, 168
follow-up communications, 171
Food and Drug Administration (FDA), 5
frequency of audits, 91. *See also*
    auditors
functional expansion or performance
    improvement, 146
functional requirements, 131
    new functionality, 132

G

Gantt chart, 31, 179, 180, 184
get started, 25, 26, 34
global database, 148
government sanction, 37, 38
guidelines for services, 8, 159

H

hand assembly, 57, 58
"HOLD" tag, 83
human factors, 169–71
    management tools, 171
human resources, 66, 174

I

identification and traceability, 71, 82,
    97, 112
IECQ/NECQ certification, 41
implementation of ISO, 27, 28, 66
    awareness, 23
    cleanup, 181
    commitment and active involvement
        of upper management, 27

coordinating, 25
core team, 27
discipline, 181
distraction, 30
do's and don'ts, 26
estimating the time and effort
    required, 24
funding, 30
how the standards work, 23
monitoring progress, 179
time factors, 26
implementation sequence, 128
import tariffs, 2, 9
improve efficiency and reduce costs, 19
improvement activity, 186
improvements in productivity, 16
in-house training personnel, 29
in-process inspection, 75, 78, 79, 98
included software product, 155
incoming inspection, 12, 57, 77, 78, 83, 98
increase customer satisfaction, 19
independent third-party audit, 177, 188
industrial output, 1, 3
ineffective implementation, 107
information sharing, 174. *See also*
    informational meetings
information society, 3, 4
informational meetings, 175
input information, 14. *See also* feedback
    information
inputs, 139
inspection, 6, 7, 12, 16, 19, 41, 46, 51,
    55, 57, 58, 65, 66, 73, 74, 76–79, 82,
    83, 85, 86, 98, 99, 104, 112, 147,
    172
    comprehensive records, 78
inspection and test status, 82, 99
inspection and testing, 7, 65, 71, 74, 76,
    77, 98
inspection records, 79
installation, 45, 59, 65, 66, 72, 121, 131,
    144, 145
installation site, 131
integration level, 142
interface modification, 146
internal audits, 20, 29, 31, 37, 49, 92,
    128, 169, 181, 185, 186, 188
    corrective action system, 186

internal customers, 14
internal process experts, 28
internal quality audits, 48, 71, 91, 93,
    100, 107, 128
International Accreditation Forum, 38
International Electrotechnical
    Commission's Quality, 41
International Organization for
    Standardization, 3, 5
  history of, 1–6
      ISO 9000 series first edition, 5
      ISO 9000 series 2nd edition, 5–6
international standard of
    measurements, 5
international standards, 65, 122, 189
interrelations among all personnel, 104
inventions (U.S.), 4
ISO 9000 certification, 6, 7. *See also* ISO
    9000 registration
  selling point, 7
ISO 9000 registration, 7, 9, 28, 41, 101, 109
  accreditation, 19
  audit by an external registration
    organization, 3
  benefits to customer, 11–15
  benefits to subcontractor, 15–18
  benefits to supplier, 18–20
  gaining momentum, 18
ISO 9000 standards
  full service business system, 106
  purchasing requirements, 112
ISO 9000-3 document, 122
ISO coordinator, 25, 31, 177, 178, 180, 184
ISO guidelines, 6, 16, 19–21, 61
  additional guidelines, 8
  five basic standards, 7
  major areas, 19
  structure, 6
ISO implementation project team, 178
"ISO overkill", 34
ISO requirements, 112, 119, 181, 184
ISO seminars, 34
ISO software-specific registration, 124
ISO standards
  how to benefit from, 189
  selecting a model, 45
  the future of standards, 189
ISO steering committee, 181

J

job descriptions, 47, 52, 135
joint reviews, 125, 126
Just in Time Manufacturing (JITM), 113
Just in Time Training (JITT), 113
Just-in-Time (JIT), 18, 114. *See also*
    pull system
  demands of JIT, 117
  electronic data interchange between
    vendors and buyers, 118
  how ISO and JIT reinforce each
    other, 118

K

Kanban, 115, 116, 119
  withdrawal Kanban, 115, 116
key parameters, 94, 95

L

leading causes for quality failures, 20
"lean and mean" approach, 27
Level I documents, 51
life cycle maintenance phase, 146
life-cycle model, 128
line supervisors, 117
linear design and development model,
    63
linear design process, 62
"lone ranger" style of management, 102
lot files, 73

M

maintenance activities, 146, 147, 155,
    157
maintenance audits, 188
maintenance phase, 146
maintenance plan, 146, 148, 157
maintenance records, 76, 147, 173
major functions, 131, 175
management involvement, 187
management representative, 48, 96, 125
management responsibility, 45, 48, 96,
    104, 106, 124, 125
  organization, 47–48
management review, 24, 27, 49, 50, 71,
    91, 93, 96, 107, 169, 176, 181, 185,
    187, 190
management tools, 136, 149

manufacturing and service cultures, 103
manufacturing hub, 74
manufacturing process, 97, 102, 118
manufacturing resource planning, 110, 119
market expansion, 42
market research, 161, 175
marketing advantage, 19
marketing and sales, 23, 56
Marshall Plan, 4, 9
master document list, 151
master production schedule, 110
material acquisitions, 110
Material Review Board (MRB), 85
Materials Requirement Planning
    (MRP), 83, 109–12
  levels of MRP systems, 110
  relationship of MRP and ISO, 111
measurement capability studies, 81
measurement equipment, 76, 79, 80, 98
  identification, 80
  measurement capability, 81
  recall system, 98
member nations, 5, 7
method of control, 51, 69, 70, 86, 173
methodologies, 127, 136, 138–40, 147, 149
military or governmental regulations, 24
mixing, 168
moderator, 141
modern design team, 59
modifications to contracts, 56
motivation efforts, 55
multi-site company, 25
multiple registration, 38, 42

N

National Accreditation Council for
    Certification Board, 37
National Electronic Components
    Quality (NECQ), 41
National Institute of Standards and
    Technology (NIST), 38, 80
new functionality, 132, 143
new technology, 13, 161, 175
non-production quality issues, 106
nonconformance, 75, 84–86, 88, 89, 178,
    179, 184, 186, 187, 190
nonconforming material, 78, 82, 84–86,
    90, 99
nonconforming process conditions, 89

nonconforming product, 19, 73–76,
    83–85, 89, 99
  evaluation, segregation, and
    disposition of, 85
  identification, 84
  notification, 86
North Atlantic Treaty Organization
    (NATO), 4

O

objectives for quality, 46, 50
objectivity checks, 181
operational constraints, 131
operations control, 72
operator error, 89, 90, 99
organizational charts, 47, 50
organizational culture change, 103
organizational overview, 126
Organization for Economic Cooperation
    and Development (OECD), 4
outside providers of warehousing and
    shipment service, 87

P

packaging, 20, 56, 65–67, 86, 87, 99,
    106, 110, 112, 144
Pareto Chart of defects, 94
part-time employees, 168
partnership method, 16
partnerships with suppliers, 89, 117, 119
PC-based software programs for project
    planning, 180
performance testing, 60
periodic re-evaluation of product, 13
periodic tests or inspection, 52
personnel motivation, 54
postwar era, 2
  Japan and Germany, 2
  United States of America, 1
pre-control, 95
preassessment, 37
preliminary audit, 182, 184
proactive measures, 126
problem resolution, 146, 147
process certification study, 76
process control, 19, 57, 71, 73–77, 79,
    93, 94, 97, 98, 174
  erroneous process decisions, 95
  variation of the process, 96

process flow chart, 29
process measurements, 75, 152, 153
process metrics, 151
process steps and quality checks, 83
process technicians, 79
processing deficiencies, 76
processing operations, 74
processing price, 57
produce wealth, 3, 9
product identification, 71, 72, 74, 82, 97, 112
product inventory, 115
product liability problems, 18
product life cycle, 121–23, 146, 148
product lines, 47, 108, 112
product metrics, 151
product nonconformities, 153
product quality measurements, 152
product reports, 76
product specification, 60
product/service continuum, 159
production equipment, 66, 74, 76
production, installation, and servicing, 7, 121
production line procedures, 122
production schedule, 74, 110
production sheets, 73
production stage, 122
production supervisors, 52, 117
professional ISO trainers, 29
professional networks, 26
programmers, 129, 132
programming rules, 139
progress review schedules, 135
progress reviews, 136
project development planning, 135. *See also* development plan
project implementation, 130
project manager, 135, 136
project teams, 28
promotional tools, 27
protocol, 60
prototype tests, 65
pull system, 113, 114
purchase order, 56, 77
purchased products, 154. *See also* third-party vendor
purchased software, 67
purchaser's representative, 133

purchaser's requirements specification, 130
purchasing, 6, 16, 18, 19, 62, 66, 89, 112, 119, 168
purchasing department, 18, 77, 89
purchasing function, 56, 89
push system, 113, 114

Q

quality assessment, 139, 169, 176
quality concerns, 123, 130, 137
quality control, 6, 83, 122, 126, 160, 164, 169, 170, 176
    quality management, 7
quality engineer, 108
quality improvement team, 54
quality loop, 14, 162
Quality Management System (QMS), 3
quality manual, 24, 26, 28, 31, 49, 50, 51, 55, 71, 126, 127, 133, 156, 173, 176
    Level I documents, 50
    record retention time, 71
    skeleton view, 50
quality manual development team, 50
quality movement, 11
quality organization, 48, 117
quality plan, 76, 127, 128, 137, 138, 156, 173
    content, 137–138
quality policy, 24, 46, 47, 50, 51, 91, 96, 118, 165, 166, 173
    generic, 46
quality records, 6, 20, 24, 50, 51, 58, 60, 71, 72, 74, 79, 82, 83, 98, 100, 107, 126, 128, 129, 135, 136, 138, 139, 151, 153, 155, 157, 173, 176, 183
quality service, 159, 165
    design changes, 164
    design responsibilities, 163
    documentation methodology, 172
    human aspects, 160
    management functions, 165
    management review, 169
    new technology, 161
    quality control design, 170
    quality objectives and activities, 165
    quality records for service providers, 173–174

quality specifications, 163
service delivery specification, 163
service specification, 163
quality systems, 4, 8, 16, 39, 40, 76, 112,
    123, 172
  computer software, 8
  services, 8
quality techniques, 101
questions and concerns, 130

R

Raad voor de Certificate (RvC), 37
recall system, 81
records and reports, 147
redeployment of internal resources, 30
reduced cost, 14, 48
referrals and free advertising, 14
Registrar Accreditation Board
    (RAB), 37
registrar audit, 37
  choosing a registrar, 40
  memoranda of understanding, 41
registrar audits
  accreditation process criteria, 38
registration audit, 25, 178, 181–85, 188
regression tests, 143
reinspection, 12, 86
release coordination, 148
release policy document, 148
removal of registered status, 187
repetitive problems, 90
replication, 130, 144, 145, 157
requirement specifications, 124, 125, 130
requirements review sessions, 132
requirements specification, 131
requirements specifications
  approving and implementing
      changes, 133
reservation systems, 167
resource requirements, 48
response time, 166, 167, 175
revalidation, 165
reviewers, 141
rework, 12, 85, 86, 94
root cause analysis, 88, 146
rules, practices, and conventions, 153

S

safety and environmental
    considerations, 65
safety and liability, 60
safety and reliability, 13
sampling customer reactions, 171, 172.
    See also follow-up communications
scientific and technological
    developments, 3
scope of maintenance, 147
scrap disposition tickets, 86
security, 67, 70, 82, 131, 145, 171
service brief, 161–63, 171, 175
service changes, 164
service delivery process, 171, 172, 174
  evaluation of, 172
service quality design, 160
service quality loop, 162
service scheduling, 167
service sector, 3, 4, 9, 24, 159, 161, 169
servicing, 7, 20, 58–60, 62, 100, 106,
    121, 122, 130, 146, 167, 169
  documentation package, 59
set points, 75
"ship as is" decision, 85
six sigma, 6
Society of Manufacturing Engineers
    (SME), 11
software developer/provider, 123
software development. *See also*
    computer software
  design phase, 138
  design reviews, 140
  implementation, 139
  software design specification, 139
  software developers, 122
software development environment,
    121, 122, 146, 154, 156
software engineering training, 124
software industry, 8, 121, 123
software quality, 121, 124, 126, 150,
    151, 153, 156
Software Quality Manual, 126
Software Quality System Registration
    (SQSR), 124
software tools, 140

source code, 127, 140, 141, 155
SPC implementation, 107
special processes, 65, 76, 77, 97, 98
specifications and service requirements, 65
split shifts, 167
staggering, 167
staging area, 113
standing management committee, 106
statistical process control, 77, 79, 94, 98, 107, 108
   trouble-shooting techniques, 108
Statistical Process Control (SPC), 94
statistical techniques, 20, 52, 54, 74, 93, 100
   commonly used statistical tools, 93
stopped line, 117
storage and shipping, 87
strategic planning, 49, 51, 105. *See also* TQM planning
strategic pricing, 167
structure of the organization, 126, 127, 173
structuring the design review process, 62. *See also* design cycle
sub-contract requirements, 15
sub-contractor records, 71
subassemblies, 72, 73, 110, 113
subcontractors, 11, 15–18, 21, 112, 118, 154, 163, 168, 173, 176, 190
   agreement with, 16
supplier, 15–20
   records of supplier's performance, 57
supplier audit, 30, 38, 77
supplier problems, 88, 89
supplier-customer partnership, 14
suppliers
   performance ratings, 77
support, 4, 18, 26, 27, 46, 48, 59, 60, 66, 73, 110, 147, 155, 163, 172
surveillance audit, 38, 72, 91, 185–90
surveys and interviews, 161. *See also* market research
suspect product, 75, 77
system level, 128, 142
system overview, 131
system weaknesses, 185
system-level acceptance test, 131

systematic structure, 103
systemic correction, 90

T

tactics based on intimidation and coercion, 18
team dynamics, 28, 55
team efforts, 11, 112, 174, 175
team tactics, 102
team training, 54
technical factors, 169, 170, 176
   control measures, 170
technical interfaces, 60
technical manager, 47, 48
technical personnel, 54
technical responsible, 136
temporary change form, 70
temporary procedure, 109
test cases, 136, 142–44
test environment, 143
test market, 60
test plan review and approval, 144
test planning, 142, 144
test records, 79. *See also* inspection records
test result validity, 82
test software, 82
tester, 136
third-party vendor, 154, 157
"throw it over the wall", 16
TickIT auditors, 123
TickIT Guide, 123
tools and techniques, 153, 154
Total Quality Control (TQC), 101
Total Quality Management (TQM), 6, 101
   chief executive officer, 105
   complementary systems, 106
   evolution of, 102
   management involvement, 104
   structure, 104
   TQM planning, 105
   TQM structure, 107
total quality subsystem, 113
traceability, 65, 71–73, 79, 82, 97, 112, 168
traceability system, 73
tracking continuous improvement activities, 84

trade deficits, 1
traditional mass production, 113–17,
    119. *See also* push system
  problems related to the inventories, 115
  quick fixes, 117
training, 6, 20, 24–26, 28–30, 36, 39, 40,
    48, 49, 51–55, 59, 60, 66, 71, 76, 90,
    92, 93, 100, 102, 104, 106, 107, 113,
    124, 127, 145, 155, 156
  how training is conducted, 52
training auditors, 92
training consultants and vendors, 53
training need assessments, 156
training program, 24, 51–55
  areas for education and training, 174
  basic statistical techniques, 52
  training objectives, 53
training sessions, 54
traveler, 83
turnover rate, 28, 29
two-party contractual situations, 123
typical audit, 74

U

uncontrolled, 70
Underwriter's Laboratories (UL)
  approval, 18

unit level, 142
universal registration, 38
universal tag, 72
up-to-date procedures, 66, 71
use-as-is processing, 86
user interface, 127, 132, 143, 150, 153

V

validation, 65, 131, 137, 142, 143, 150,
    154, 155, 163, 164, 175
validation of purchased product, 154
value and quality, 2
  quality climate, 14
  quality loop, 12
verification activities, 77
verification procedures, 59, 127, 134

W

walk-throughs, 140, 155. *See also* code
  inspections
work flow, 115
work instructions, 51, 69, 71, 74, 88
work instructions and procedures, 51, 69
work-in-progress (WIP), 112
written procedure, 28, 29, 31, 51, 58, 67,
    74, 75